Recent volumes in MSJ Memoirs

1 **Ivan Cherednik, Peter J. Forrester and Denis Uglov**, Quantum many-body problems and representation theory, 1998
2 **Masako Takahashi, Mitsuhiro Okada and Mariangiola Dezani-Ciancaglini (Eds.)**, Theories of types and proofs, 1998
3 **Tomotada Ohtsuki**, Combinatorial quantum method in 3-dimensional topology, 1999
4 **Atsushi Matsuo and Kiyokazu Nagatomo**, Axioms for a vertex algebra and the locality of quantum fields, 1999
5 **Daryl Cooper, Craig D. Hodgson and Steven P. Kerckhoff**, Three-dimensional orbifolds and cone-manifolds, 2000
6 **Kong De-xing**, Cauchy problem for quasilinear hyperbolic systems, 2000
7 **Vladimir Georgiev**, Semilinear hyperbolic equations, 2000
8 **Eric M. Opdam**, Lecture notes on Dunkl operators for real and complex reflection groups, 2000
9 **Peter Orlik and Hiroaki Terao**, Arrangements and hyper-geometric integrals, 2001
10 **Yuri G. Prokhorov**, Lectures on complements on log surfaces, 2001
11 **John R. Stembridge, Jean-Yves Thibon and Marc A. A. van Leeuwen**, Interaction of combinatorics and representation theory, 2001
12 **Yves André**, Period mappings and differential equations. From $\mathbb{C}$ to $\mathbb{C}_p$, 2003
13 **Shigeaki Koike**, A beginner's guide to the theory of viscosity solutions, 2004
14 **Noboru Nakayama**, Zariski-decomposition and Abundance, 2004
15 **Valery Alexeev and Viacheslav V. Nikulin**, Del Pezzo and K3 surfaces, 2006
16 **Brian H. Bowditch**, A course on geometric group theory, 2006
17 **Arkady Berenstein, David Kazhdan, Cédric Lecouvey, Masato Okado, Anne Schilling, Taichiro Takagi and Alexander Veselov**, Combinatorial Aspect of Integrable Systems, 2007
18 **Yasutaka Ihara**, On Congruence Monodromy Problems, 2008
19 **Joseph Najnudel, Bernard Roynette and Marc Yor**, A Global View of Brownian Penalisations, 2009
20 **Danny Calegari**, scl (Stable Commutator Length), 2009
21 **Gautami Bhowmik, Kohji Matsumoto and Hirofumi Tsumura (Eds.)**, Algebraic and Analytic Aspects of Zeta Functions and L-functions, 2010
22 **Michael Ruzhansky and James Smith**, Dispersive and Strichartz estimates for hyperbolic equations with constant coefficients, 2010
23 **Armen Sergeev**, Kähler geometry of loop spaces, 2010
24 **Taro Asuke**, Godbillon–Vey class of transversely holomorphic foliations, 2010
25 **Hiroshi Sugita**, Monte Carlo method, random number, and pseudorandom number, 2011
26 **Shinya Nishibata and Masahiro Suzuki**, Hierarchy of semiconductor equations: relaxation limits with initial layers for large initial data, 2011

27 **Suhyoung Choi**, Geometric structures on 2-orbifolds: Exploration of discrete symmetry, 2012

28 **Toshio Oshima**, Fractional calculus of Weyl algebra and Fuchsian differential equations, 2012

29 **Takeshi Hirai, Akihito Hora and Etsuko Hirai**, Projective representations and spin characters of complex reflection groups $G(m, p, n)$ and $G(m, p, \infty)$, 2013

30 **Tatsuo Nishitani**, Cauchy Problem for Noneffectively Hyperbolic Operators, 2013

31 **Satoshi Takanobu**, Bohr–Jessen Limit Theorem, Revisited, 2013

32 **Hiroshi Isozaki and Yaroslav Kurylev**, Introduction to spectral theory and inverse problem on asymptotically hyperbolic manifolds, 2014

33 **Masaki Maruyama**, Moduli spaces of stable sheaves on schemes, 2016

34 **Martin T. Barlow, Tibor Jordán and Andrzej Zuk**, Discrete Geometric Analysis, 2016

35 **Osamu Fujino**, Foundations of the minimal model program, 2017

36 **Soichiro Katayama**, Global solutions and the asymptotic behavior for nonlinear wave equations with small initial data, 2018

37 **Kazuki Hiroe, Hiroshi Kawakami, Akane Nakamura and Hidetaka Sakai**, 4-dimensional Painlevé-type equations, 2018

38 **Alex Casella, Dominic Tate and Stephan Tillmann**, Moduli spaces of real projective structures on surfaces, 2020

39 **Masaharu Taniguchi**, Traveling Front Solutions in Reaction-Diffusion Equations, 2021

40 **Yuji Odaka and Yoshiki Oshima**, Collapsing K3 Surfaces, Tropical Geometry and Moduli Compactifications of Satake, Morgan–Shalen Type, 2021

41 **Tomoki Nakanishi**, Cluster Algebras and Scattering Diagrams, 2023

42 **Osamu Fujino**, Cone and contraction theorem for projective morphisms between complex analytic spaces, 2024

43 **Naofumi Honda and Luca Prelli**, Multi-normal deformation and multi-specialization, 2025

Multi-normal deformation and multi-specialization

Naofumi Honda and Luca Prelli

MSJ Memoirs

Mathematical Society of Japan

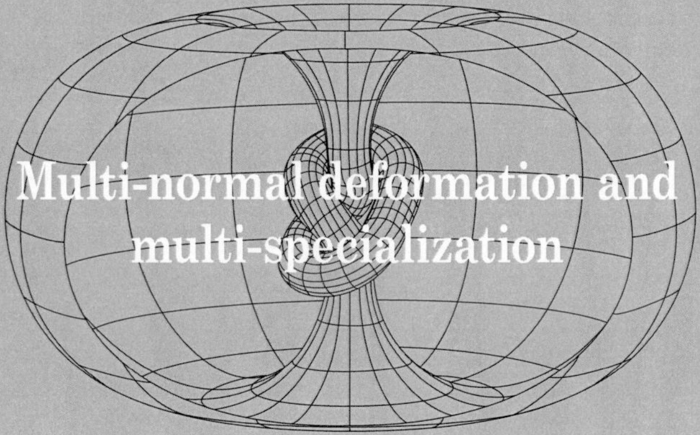

Multi-normal deformation and multi-specialization

Naofumi Honda and Luca Prelli

Volume 43 2025

Authors
NAOFUMI HONDA
Department of Mathematics, Faculty of Science, Hokkaido University,
Sapporo 060-0810, Japan.
n-honda@math.sci.hokudai.ac.jp

LUCA PRELLI
Dipartimento di Matematica, Università degli Studi di Padova,
Via Trieste 63, 35121 Padova, Italia.
lprelli@math.unipd.it

Keywords: Algebraic analysis, Asymptotic expansions, Specialization, Normal deformation, Subanalytic sheaves.

2020 Mathematics Subject Classifications: Primary 32C35, Secondary 35A27.

MSJ Memoirs

This monograph series is intended to publish lecture notes, graduate textbooks and long research papers* in pure and applied mathematics. Each volume should be an integrated monograph. Proceedings of conferences or collections of independent papers are not accepted.

Articles for the series can be submitted to one of the editors in the form of hard copy. When the article is accepted, the author(s) is (are) requested to send a camera-ready manuscript.

* long research papers are limited to contributions by MSJ members

Editorial Board

The series *MSJ Memoirs* is published occasionally (3–5 volumes each year) by the Mathematical Society of Japan. The titles and prices of recent volumes are listed on page i of this volume. Each volume of the series can be separately purchased by placing an order with:

- The Mathematical Society of Japan, 1-34-8 Taito, Taito-ku, 110-0016 Tokyo, Japan (from inside Japan)
- World Scientific Publishing, 5 Thoh Tuck Link, Singapore 596224 (from outside Japan)

Preface

Asymptotically developable expansions of holomorphic functions on a sector are an important tool to study ordinary differential equations with irregular singularities. In higher dimension H. Majima introduced in [18] the notion of strongly asymptotically developable functions along a normal crossing divisor. These functions are related with Whitney holomorphic functions on a multi-sector, as proven in [7]. The aim of this paper is to perform the generalization and the functorial construction of asymptotics, with the introduction of the notion of multi-asymptotic expansions. New geometrical tools will be required, in particular the multi-normal deformation associated with an action and a family of submanifolds. Locally we can construct new sheaves of multi-asymptotically developable functions closely related with asymptotics along a subvariety with a simple singularity such as a cusp. These geometrical and functorial constructions extend the previous ones from [24], [25], [5], [2], [9] and [8].

We have established, in [9] and [8], the theory of multi-specialization and multi-microlocalization along a simultaneously linearizable family χ of closed submanifolds (see Definition 1.1 and Proposition 1.2) on which we additionally impose some moderate condition for their geometrical configuration, that is, the condition H2 of [9] for χ. A typical example of such a configuration (with the condition H2) of two submanifolds is that they transversally intersect or that one submanifold completely contains the other. However, some important cases do not satisfy this geometrical condition such as two submanifolds which cleanly intersect but not transversally or a simultaneously linearizable family of Lagrangian submanifolds in a cotangent bundle, where we encounter the latter situation in the micro-support estimation of a multi-microlocalized object (Theorem 3.6 [8]).

The primary purpose of our paper is to give a new description of the geometry appearing in the multi-specialization along a general linearizable family χ (without the condition H2), and then, to extend several properties of the multi-specialization functor already established in [9] to this general χ. One should be aware that, however, the geometry appearing in the multi-specialization for a general χ is quite different from that for a χ with the condition H2. For a usual specialization along a closed submanifold M, a (locally defined) conic cone along the submanifold M can be considered as a fundamental geometrical object of this specialization in the sense that both the normal cone $C_M(\cdot)$ along M and a stalk formula for the specialization functor $\nu_M(\cdot)$ along M can be described by using a family of these cones. When we consider the multi-specialization along a simultaneously linearizable family χ

of submanifolds $\{M_1, \ldots, M_\ell\}$ in X with the condition H2, the situation is quite similar as that for the usual specialization. The fundamental geometrical object in this case is a multi-cone that is, by definition, the intersection of ℓ conic cones along submanifolds $M_1, M_2, \ldots, M_\ell$, respectively. And the multi-normal cone $C_\chi(\cdot)$ along χ and a stalk formula for the multi-specialization functor $\nu_\chi(\cdot)$ along χ can be described by using a family of these multi-cones as we have shown in [9].

On the other hand a multi-cone for a general linearizable family χ can be no longer obtained by a set-theoretical operation of conic cones along submanifolds. In fact, as we see in Section 2, it is defined by a semigroup generated by monomials associated with local actions μ_k's ($k = 1, \ldots, \ell$), where each action μ_k is induced from the one on the conormal bundle $T_{M_k}X$ of M_k with a local identification $X \simeq T_{M_k}X$ near M_k. Furthermore, when we drop the condition H2, we cannot either expect the zero section S_χ of the multi-normal deformation $\widetilde{X}_\chi$ along χ, i.e., the subset $\{t = 0\}$ in $\widetilde{X}_\chi$, to be a vector bundle over the base space $M = M_1 \cap \cdots \cap M_\ell$. In fact, let us consider χ which consists of two submanifolds intersecting cleanly but not transversally. Then, as Example 1.15 shows, S_χ is not a vector bundle over M even for this simple case.

Several unexpected facts are observed in geometry of the multi-specialization for a general χ as we have already seen, however, Theorems 2.31 and 3.8 and Equation (3.2) in this paper say that, by using our newly introduced multi-cones, we can characterize the multi-normal cone and establish a stalk formula for the multi-specialization functor in a usual way. This is the main goal of Section 3 in which we define the functor of multi-specialization extending the construction of [9]. We also study functorial operations of multi-specializations such as a direct image and an inverse image in our general settings. By an operation associated with a desingularization map, for example, we can show isomorphisms between several solution sheaves of holomorphic functions multi-asymptotically developable along different kinds of families of closed complex submanifolds. As such an application to $\mathcal{D}$-modules, in the forthcoming paper [12], we establish direct and inverse image formulas for the multi-specialization of the sheaf of Whitney holomorphic functions and that of tempered holomorphic functions. In particular, we deeply study direct image formulas to specific but important class of maps like a multi-conic map and a desingularization map which also appears in [4] as an important geometrical situation.

The shape of a multi-cone becomes a little bit complicated for a general χ, however, it still has many good properties. It is, in particular, infinitesimally stable under the actions μ_k's. This fact suggests us that an asymptotic expansion can be defined on this multi-cone since it is, roughly speaking, nothing but a formal Taylor expansion along orbits generated by the actions μ_k's. As a matter of fact, in Section 4, the notion of multi-asymptotic expansions along χ introduced in [9] is successfully extended to the one for a general simultaneously linearizable family χ. Further, our generalization of the multi-asymptotic expansion can manipulate the case that each action μ_k is not necessarily linear. Hence, by choosing a family of appropriate non-linear actions, locally we can construct a new sheaf of multi-asymptotically developable functions which is closely tied with asymptotics along a subvariety with a simple singularity such as a cusp.

We study, in Section 5, the functorial nature of multi-asymptotics, proving some vanishing theorems which provide a general Borel-Ritt exact sequence in this setting. Similar vanishing results hold for the multi-specialization of tempered holomorphic functions. Examples of our previous constructions are given in Section 6, in particular in $\mathbb{C}^2$ and $\mathbb{C}^3$. We also give, in the two manifolds case, a classification for multi-specializations provided by the matrix associated with the multi-normal deformation. We end this work with an Appendix in which we introduce the notion of multi-conic sheaves. Using o-minimal geometry we also obtain a decomposition theorem for subanalytic open sets needed for study of the geometry of sections of multi-conic sheaves.

At the end of the introduction, we clarify a relation between this paper and the unpublished paper [10]. As the latter paper is quite long, the authors have decided to divide the latter one into two parts, the functorial aspect of general multi-specializations and an application to multi-asymptotics. This paper is nothing but the first part, where we study functorial construction of multi-specializations and their properties from a functorial point of view. The second part, which will be published as the paper [11], is devoted to establishment of the theory of multi-asymptotics under more general geometry considered in this paper.

For this work, the first author is supported in part by JSPS KAKENHI Grant 21K03284, and the second author is supported in part by Gruppo Nazionale per l'Analisi Matematica, la Probabilità e loro Applicazioni GNAMPA - INdAM. We would like to express our thanks to these supports.

Naofumi Honda and Luca Prelli
Sapporo and Padova,
June 27th, 2024.

Contents

1 Multi-normal deformation

In [9] the notion of multi-normal deformation was introduced. Here we consider a generalization where we replace the condition H2 with a weaker one. We refer to [14] for the classical normal deformation with respect to one submanifold.

1.1 A simultaneously linearizable family of manifolds

Let X be a real analytic manifold with $\dim X = n$. Hereafter all the manifolds appearing in this paper are assumed to be countable at infinity. Let $\chi = \{M_1, \dots, M_\ell\}$ be a family of closed real analytic submanifolds of X. Set $M := M_1 \cap \cdots \cap M_\ell$. Throughout this note, we always assume that all the submanifolds are connected and that M is also connected. Recall that, for closed submanifolds M and M', we say that M and M' *intersect cleanly* at $q \in M \cap M'$ if and only if $M \cap M'$ is a closed submanifold near q and

$$(T^*_M X)_q + (T^*_{M'} X)_q = (T^*_{M \cap M'} X)_q \tag{1.1}$$

holds.

Definition 1.1. *We say that χ is simultaneously linearizable at $q \in M$ if M_j and M_k $(1 \le j, k \le \ell)$ intersect cleanly at q and if there exist a vector subspace V of $(T^*X)_q$ and its decomposition $V = V_1 \oplus \cdots \oplus V_m$ by vector subspaces such that every $(T^*_{M_j} X)_q$ is a direct sum of some V_k's.*

Note that if χ satisfies the condition H2 of [9], then it is simultaneously linearizable.

Proposition 1.2. *Let $\chi = \{M_1, \dots, M_\ell\}$ be a family of closed submanifolds of X and $q \in M$. Then the following conditions are equivalent.*

i. χ is simultaneously linearizable at q.

ii. There exist a neighborhood W of q, a neighborhood U of 0 with a system of local coordinates $(x_1, \dots, x_n)$, the isomorphism $\psi : W \to U$ and subsets I_j's of $\{1, \dots, n\}$ $(j = 1, \dots, \ell)$ such that each $\psi(M_j \cap W)$ is defined by equations $x_i = 0$ $(i \in I_j)$.

The proof of the proposition is rather long, and hence it is given in Appendix A.4. Note that the decomposition $V = V_1 \oplus \cdots \oplus V_m$ appearing in Definition 1.1 is not unique. However, if we consider minimal decomposition such that Definition 1.1 holds, the number of V_k's of such a decomposition is unique, and this minimal number m is called *rank of the family χ at q*. Furthermore, when χ is simultaneously linearizable at any point in M, as M is connected by assumption, the rank of the family χ does not depend on the point under consideration.

By Proposition 1.2, we have locally the family $\{I_j\}_{j=1}^{\ell}$ of subsets in $\{1, 2, \ldots, n\}$ so that each M_j is defined by $x_k = 0$ for $k \in I_j$. Let us consider the family $\{\widehat{I}_j\}_{j=1}^{m}$ of equivalence classes of $\bigcup_{j=1}^{\ell} I_j$ by the equivalence relation

$$i_1 \sim i_2 \iff \text{"} i_1 \in I_j \iff i_2 \in I_j \text{ for all } j = 1, 2, \ldots, \ell\text{"}. \tag{1.2}$$

Note that the number m of equivalence classes coincides with the the rank of the family χ. For convenience, we set

$$\widehat{I}_0 = I_0 := \{1, \ldots, n\} \setminus (\bigcup_{j=1}^{\ell} I_j). \tag{1.3}$$

1.2 Geometry of multi-normal deformation

First recall the classical construction of [14] of the normal deformation of X along M_1. We denote it by $\widetilde{X}_{M_1}$ and we denote by $t_1 \in \mathbb{R}$ the deformation parameter. Let $\Omega_{M_1} = \{t_1 > 0\}$ and let us identify $s^{-1}(0)$ with $T_{M_1}X$. We have the commutative diagram

$$\begin{array}{ccccc} T_{M_1}X & \xrightarrow{s_{M_1}} & \widetilde{X}_{M_1} & \xleftarrow{i_{\Omega_{M_1}}} & \Omega_{M_1} \\ \downarrow{\scriptstyle \tau_{M_1}} & & \downarrow{\scriptstyle p_{M_1}} & \swarrow{\scriptstyle \widetilde{p}_{M_1}} & \\ M & \xrightarrow{i_{M_1}} & X. & & \end{array} \tag{1.4}$$

Set $\widetilde{\Omega}_{M_1} = \{(x; t_1) \ ; \ t_1 \neq 0\}$. Let $\chi = \{M_1, \ldots, M_\ell\}$ be the simultaneously linearizable family given in the previous subsection, and define

$$\widetilde{M_2} := \overline{(p_{M_1}|_{\widetilde{\Omega}_{M_1}})^{-1} M_2}.$$

Then $\widetilde{M_2}$ is a closed smooth submanifold of $\widetilde{X}_{M_1}$ as χ is simultaneously linearizable.

Now we can define the normal deformation along M_1, M_2 as

$$\widetilde{X}_{M_1,M_2} := (\widetilde{X}_{M_1})_{\widetilde{M_2}}^{\sim}.$$

Then we can define recursively the normal deformation along χ as

$$\widetilde{X} = \widetilde{X}_{M_1,\ldots,M_\ell} := (\widetilde{X}_{M_1,\ldots,M_{\ell-1}})_{\widetilde{M_\ell}}^{\sim}.$$

Set $S_\chi = \{t_1, \ldots, t_\ell = 0\}$, $M = \bigcap_{i=1}^{\ell} M_i$ and $\Omega = \{t_1, \ldots, t_\ell > 0\}$. Then we have the commutative diagram

$$\begin{array}{ccccc} S_\chi & \xrightarrow{s} & \widetilde{X} & \xleftarrow{i_\Omega} & \Omega \\ \downarrow{\scriptstyle \tau} & & \downarrow{\scriptstyle p} & \swarrow{\scriptstyle \widetilde{p}} & \\ M & \xrightarrow{i_M} & X. & & \end{array} \tag{1.5}$$

In what follows, we always assume:

A.1 χ is a simultaneously linearizable family of manifolds at any point in M.

We denote $\mu_\chi : \widetilde{X} \times (\mathbb{R}^+)^\ell \to \widetilde{X}$ the multi-action associated with χ. Note that, as $\mu_\chi(S_\chi, \lambda) \subset S_\chi$ holds for $\lambda \in (\mathbb{R}^+)^\ell$, μ_χ induces the action on S_χ, which is denoted by the same symbol hereafter. Then, for a generic point $p \in S_\chi$, the orbit $\mu_\chi(p, (\mathbb{R}^+)^\ell) \subset S_\chi$ becomes a real analytic manifold, whose maximal dimension is called *the dimension of an orbit of* μ_χ.

Definition 1.3. *We say that the multi-action μ_χ has maximal rank if the dimension of an orbit of μ_χ is equal to either the number of submanifolds in χ (usually denoted by ℓ) or the rank of the family χ (usually denoted by m). In particular, μ_χ is said to be non-degenerate if the dimension of the orbit of μ_χ coincides with ℓ and transitive if it coincides with m.*

Then we have:

Lemma 1.4. *The following conditions are equivalent.*

1. *The multi-action μ_χ is non-degenerate.*
2. *$\mu_\chi(x, \lambda) = \mu_\chi(x, \lambda')$ for any $x \in S_\chi$ implies $\lambda = \lambda'$.*

Now we give the local description of the multi-normal deformation if χ satisfies the condition A.1. Let $\{I_j\}_{j=1}^\ell$ be the family of subsets in $\{1, \dots, n\}$ in Proposition 1.2, and let $\{\widehat{I}_k\}_{k=1}^m$ be the equivalence classes in $\bigcup_{j=1}^\ell I_j$ by the equivalence relation (1.2), where m coincides with the rank of the family χ. We define the monomial φ_k $(k = 1, \dots, m)$ of the variables $t_1, \dots, t_\ell$ by

$$\varphi_k(t_1, \dots, t_\ell) = \prod_{\{j \in \{1,\dots,\ell\};\, \widehat{I}_k \subset I_j\}} t_j. \tag{1.6}$$

We also define the $\ell \times m$-matrix $A_\chi = (a_{jk})$ with

$$a_{jk} = \begin{cases} 1 & \widehat{I}_k \subset I_j, \\ 0 & \text{otherwise}. \end{cases}$$

Then we have, for $t = (t_1, \dots, t_\ell) \in (\mathbb{R}^+)^\ell$,

$$(\log \varphi_1, \dots, \log \varphi_m) = (\log t_1, \dots, \log t_\ell) A_\chi.$$

Using these φ_k's, the projection $p : \widetilde{X} \to X$ and the action μ_χ on $\widetilde{X}$ are locally defined as follows: We take

$$(x^{(0)}, x^{(1)}, \dots, x^{(m)}) \quad \text{and} \quad (x^{(0)}, x^{(1)}, \dots, x^{(m)}; t_1, \dots, t_\ell) \tag{1.7}$$

the local coordinates of X and those of $\widetilde{X}$, respectively. Here $x^{(j)}$ $(j > 0)$ denotes the set of coordinates x_i's $(i \in \widehat{I}_j)$ and $x^{(0)}$ denotes the rest of coordinates.

Then the projection p is given by, for $(x;\, t) = (x^{(0)},\, x^{(1)},\, \dots, x^{(m)};\, t_1, \dots, t_\ell) \in \widetilde{X}$,

$$p(x;\, t) = \left(x^{(0)},\, \varphi_1(t)x^{(1)},\, \dots,\, \varphi_m(t)x^{(m)}\right).$$

The action μ_χ on $\widetilde{X}$ is defined locally by, for $\lambda = (\lambda_1, \dots, \lambda_\ell) \in (\mathbb{R}^+)^\ell$,

$$\mu_\chi((x;\, t), \lambda) = \left(x^{(0)},\, \varphi_1(\lambda)x^{(1)},\, \dots,\, \varphi_m(\lambda)x^{(m)};\, t_1/\lambda_1,\, \dots,\, t_\ell/\lambda_\ell\right). \tag{1.8}$$

It is the composition of the actions μ_j associated with the I_j's, $j = 1, \dots, \ell$, which are defined locally by, for $\lambda_j \in \mathbb{R}^+$,

$$\mu_j((x;\, t), \lambda_j) = \left(x^{(0)},\, \lambda_j^{a_{j1}}x^{(1)},\, \dots,\, \lambda_j^{a_{jm}}x^{(m)};\, t_1,\, \dots,\, t_j/\lambda_j,\, \dots,\, t_\ell\right). \tag{1.9}$$

From these local observations, the following lemma easily follows:

Lemma 1.5. *μ_χ has maximal rank if and only if A_χ has maximal rank, that is,*

$$\operatorname{Rank} A_\chi = \min\{\ell,\, m\}.$$

Furthermore, μ_χ is non-degenerate if $\operatorname{Rank} A_\chi = \ell$, *and it is transitive if* $\operatorname{Rank} A_\chi = m$.

Now we introduce an important geometrical object.

Definition 1.6. *Let Z be a subset of X. The multi-normal cone to Z along χ is the set $C_\chi(Z) = \overline{\widetilde{p}^{-1}(Z)} \cap S_\chi$.*

The following result has been proved in [8]

Lemma 1.7. *Let $p = (p^{(0)}, p^{(1)}, \dots, p^{(m)};\, 0, \dots, 0) \in S_\chi$ and let $Z \subset X$. The following conditions are equivalent:*

1. *$p \in C_\chi(Z)$.*
2. *There exist two sequences*

$$\{c_\kappa = (c_{1,\kappa}, \dots, c_{\ell,\kappa})\}_{\kappa\in\mathbb{N}} \subset (\mathbb{R}^+)^\ell, \quad \left\{\left(q_\kappa^{(0)}, q_\kappa^{(1)}, \dots, q_\kappa^{(m)}\right)\right\}_{\kappa\in\mathbb{N}} \subset Z$$

such that $\varphi_j(c_\kappa)q_\kappa^{(j)} \to p^{(j)}$ $(j = 1, \dots, m)$ and $c_{j,\kappa} \to +\infty$ $(j = 1, \dots, \ell)$ when $\kappa \to \infty$.

1.3 A simultaneously linearizable family of transitive type

We first study the case where the zero section S_χ of the multi-normal deformation $\widetilde{X}$ along χ becomes a vector bundle over the base space $M := M_1 \cap \cdots \cap M_\ell$.

Definition 1.8. *We say that a simultaneously linearizable family χ is of transitive type if the associated multi-action μ_χ is transitive. Furthermore, it is said to be of normal type if μ_χ is non-degenerate and transitive.*

Let m be the rank of the family χ, and let ℓ be the number of submanifolds in χ. Note that χ is of normal type if and only if, under the local description of the multi-normal deformation given in the previous subsection, the A_χ is a square matrix, i.e., $\ell = m$ and it is invertible. Furthermore, it follows from the definition that $\ell \geq m$ holds when χ is of transitive type.

Theorem 1.9. *Assume χ is of transitive type. Then S_χ becomes a vector bundle over M.*

Proof. Suppose that X has a system of local coordinates blocks

$$(x^{(0)}, x^{(1)}, \dots, x^{(m)})$$

described in the previous subsection, where m is the rank of the family χ. Let

$$f(x) = (f^{(0)}(x),\, f^{(1)}(x), \dots, f^{(m)}(x))$$

be a coordinate transformation on X which sends each M_j to M_j $(j = 1, \dots, \ell)$. Let, for $k \geq 1$,

$$f^{(k)}(x) = \sum_{\alpha \in \mathbb{Z}^n_{\geq 0}} c^{(k)}_\alpha x^\alpha \qquad (c^{(k)}_\alpha \in \mathbb{R}^{\#\widehat{I}_k})$$

be the Taylor expansion of $f^{(k)}$ at the origin and set $f^{(k)}_\alpha(x) := c^{(k)}_\alpha x^\alpha$.

We first show the claim that

$$f^{(k)}_\alpha(x^{(0)}, \varphi_1(t)x^{(1)}, \dots, \varphi_m(t)x^{(m)}) \text{ is divided by the monomial } \varphi_k(t). \tag{1.10}$$

Let α with $c^{(k)}_\alpha \neq 0$. By definition, $f^{(k)}(x)$ vanishes on M_j for j with $\widehat{I}_k \subset I_j$, and hence, $f^{(k)}_\alpha(x)$ also vanishes on M_j for such a j, which entails that there exists $i \in I_j$ with $\alpha_i > 0$. Hence $f^{(k)}_\alpha(x^{(0)}, \varphi_1(t)x^{(1)}, \dots, \varphi_m(t)^{(m)})$ has the factor t_j for j with $\widehat{I}_k \subset I_j$, from which the claim follows.

It follows from the claim (1.10) that the induced map $\widetilde{f}$ of f on $\widetilde{X}$ is well-defined since their definitions are

$$\tilde{f}^{(k)}(x) = \frac{1}{\varphi_k(t)} f(x^{(0)}, \varphi_1(t)x^{(1)}, \dots, \varphi_m(t)^{(m)})\Big|_{t=0} \qquad (k = 1, 2, \dots, m),$$
$$\tilde{f}^{(0)}(x) = f(x^{(0)}, 0, \dots, 0),$$

and $t_j = t_j$. Now assume that there exist some $k \in \{1, \dots, m\}$ and $\alpha \in \mathbb{Z}^n_{\geq 0}$ satisfying

$$\varphi_k(t) = \prod_{j=1}^m (\varphi_j(t))^{\alpha^{(j)}},$$

where we set $\alpha^{(j)} = \sum_{i \in \widehat{I_j}} \alpha_i$. Let e_k be the unit vector with the k-th element being 1. The above equation implies

$$A_\chi e_k = A_\chi {}^t(\alpha^{(1)}, \ldots, \alpha^{(m)})$$

which is equal to

$$A_\chi({}^t(\alpha^{(1)}, \ldots, \alpha^{(m)}) - e_k) = 0.$$

Since $\operatorname{Rank} A_\chi = m$ holds, we have $({}^t(\alpha^{(1)}, \ldots, \alpha^{(m)}) - e_k) = 0$, that is, allowable α's are multi-indices in which the only one element α_i $(i \in \widehat{I_k})$ is equal to 1. Therefore the associated map with f on $\widetilde{X}$ is the one of vector bundles with the fiber coordinates $(x^{(1)}, \ldots, x^{(m)})$ whose transformation law is given by the block diagonal map

$$\frac{\partial f^{(1)}}{\partial x^{(1)}}(x^{(0)}, 0, \ldots, 0) \oplus \cdots \oplus \frac{\partial f^{(m)}}{\partial x^{(m)}}(x^{(0)}, 0, \ldots, 0).$$

Here $\dfrac{\partial f^{(k)}}{\partial x^{(k)}}$ denotes the Jacobian matrix of the map $f^{(k)}$ with respect to variables x_i's $(i \in \widehat{I_k})$. □

We continue to assume χ to be of transitive type. We now give an explicit description of S_χ. We first define m subsets $B_1, \ldots, B_m$ of $\{1, \ldots, \ell\}$ in the following way. First take a point $p \in M$ and take a system of local coordinates blocks $(x^{(0)}, x^{(1)}, \ldots, x^{(m)})$ of X given in the previous subsection, where m is the rank of the family χ. Then we set, for $1 \le k \le m$,

$$B_k := \{j \in \{1, \ldots, \ell\};\ \widehat{I_k} \subset I_j\}.$$

Since M is connected, a different choice of a point p and that of coordinates gives the same family of subsets B_k's up to permutation. For each $k = 1, \ldots, m$ introduce $T_k \subseteq \{1, \ldots, \ell\}$ saying that $j \notin T_k$ if $\exists i \in B_k$ such that M_j is transversal to M_i (locally $I_j \cap I_i$ is empty). That is,

$$T_k = \{j \in \{1, \ldots, \ell\};\ I_j \cap I_i \neq \emptyset \text{ for any } i \in B_k\}.$$

Then we have:

Proposition 1.10. *For $k = 1, 2, \ldots, m$, set*

$$N_k := \begin{cases} X & \text{if } B_k = T_k, \\ \displaystyle\bigcap_{j \in T_k \setminus B_k} M_j & \text{otherwise.} \end{cases}$$

Then S_χ is a direct sum of the following vector bundles over M:

$$\frac{TN_k \underset{X}{\times} M}{\displaystyle\sum_{j \in B_k} \left(T(M_j \cap N_k) \underset{X}{\times} M \right) + TM} \qquad (k = 1, \ldots, m).$$

Proof. Let $(x^{(0)}, x^{(1)}, \dots, x^{(m)})$ be a system of local coordinates blocks given in the previous subsection, and let $(x^{(0)}; \xi^{(1)}, \dots, \xi^{(m)})$ be that of S_χ. Then the local coordinates of $TN_k \underset{X}{\times} M$ is given by

$$(x^{(0)}; \xi^{(k')})_{k' \in \Xi}$$

where Ξ consists of indices k''s such that $\widehat{I}_{k'} \cap I_j = \emptyset$ holds for any $j \in T_k \setminus B_k$. Note that $\widehat{I}_k \cap I_j = \emptyset$ also holds for any $j \notin B_k$.

Let $k' \in \Xi$ with $k' \neq k$. Then, we can find $j \in B_k$ with $\widehat{I}_{k'} \cap I_j = \emptyset$. As a matter of fact, if such a j does not exist, then for any $i \in B_k$, we have $\widehat{I}_{k'} \subset I_i$. Since $\widehat{I}_k$ and $\widehat{I}_{k'}$ are different equivalence classes, it follows from the definition of the equivalence relation that there exists $j' \in \{1, \dots, \ell\}$ such that

$$\widehat{I}_{k'} \subset I_{j'}, \quad \widehat{I}_k \cap I_{j'} = \emptyset.$$

This implies $j' \in T_k \setminus B_k$ because of $\widehat{I}_{k'} \subset I_{j'} \cap I_i$ for any $i \in B_k$. Hence we have obtained that $T_k \setminus B_k$ is non-empty and k' does not belong to Ξ due to $\widehat{I}_{k'} \subset I_{j'}$, which contradicts the choice $k' \in \Xi$.

Therefore we conclude that the local coordinates of

$$\frac{TN_k \underset{X}{\times} M}{\sum_{j \in B_k} \left(T(M_j \cap N_k) \underset{X}{\times} M \right) + TM}$$

are given by $(x^{(0)}; \xi^{(k)})$. Now let f be a coordinate transformation considered in the proof of Theorem 1.9. Then the corresponding coordinate transformation of the above bundle is clearly given by the matrix $\dfrac{\partial f^{(k)}}{\partial x^{(k)}}(x^{(0)}, 0, \dots, 0)$, and that of S_χ is given by

$$\frac{\partial f^{(1)}}{\partial x^{(1)}}(x^{(0)}, 0, \dots, 0) \oplus \cdots \oplus \frac{\partial f^{(m)}}{\partial x^{(m)}}(x^{(0)}, 0, \dots, 0)$$

as in the proof of Theorem 1.9. Hence the bundle associated with B_k is a subbundle of S_χ over M. This completes the proof. □

We give some examples.

Example 1.11. *Let $X = \mathbb{R}^n$ with coordinates $(x^{(0)}, x^{(1)}, x^{(2)}, x^{(3)})$. Define $M_1 = \{x^{(1)} = x^{(3)} = 0\}$, $M_2 = \{x^{(2)} = x^{(3)} = 0\}$ and $M_3 = \{x^{(1)} = x^{(2)} = x^{(3)} = 0\}$. Then*

$$S_\chi = T_M M_2 \oplus T_M M_1 \oplus \frac{TX \underset{X}{\times} M}{TM_1 \underset{X}{\times} M + TM_2 \underset{X}{\times} M}.$$

Example 1.12. *Let $X = \mathbb{R}^n$ with coordinates $(x^{(0)}, x^{(1)}, x^{(2)}, x^{(3)})$. Define $M_1 = \{x^{(2)} = x^{(3)} = 0\}$, $M_2 = \{x^{(1)} = x^{(3)} = 0\}$ and $M_3 = \{x^{(1)} = x^{(2)} = 0\}$. Then*

$$S_\chi = T_M M_1 \oplus T_M M_2 \oplus T_M M_3.$$

Example 1.13. *Let $X = \mathbb{R}^n$ with coordinates $(x^{(0)}, x^{(1)}, x^{(2)}, x^{(3)})$. Define $M_1 = \{x^{(1)} = x^{(2)} = 0\}$, $M_2 = \{x^{(2)} = x^{(3)} = 0\}$ and $M_3 = \{x^{(3)} = 0\}$. Then*

$$S_\chi = T_M M_2 \oplus \frac{TX \underset{X}{\times} M}{TM_1 \underset{X}{\times} M + TM_2 \underset{X}{\times} M} \oplus T_{M_3} X.$$

Remark 1.14. *When χ is of transitive type, the zero section S_χ becomes a vector bundle over M as we have already seen. However, in general, the simultaneously linearizable condition is not enough to assure the existence of a vector bundle structure on S_χ. Another important exceptional case where S_χ has a vector bundle structure has been studied in [8].*

If χ is not of transitive type, generally S_χ is not a vector bundle over $M = \bigcap_k M_k$ as the following example shows:

Example 1.15. *Let $X = \mathbb{R}^3$ with coordinates (x_1, x_2, x_3), and let $\chi = \{M_1, M_2\}$ be a family of closed submanifolds in X defined by $M_1 = \{x_2 = x_3 = 0\}$ and $M_2 = \{x_1 = x_3 = 0\}$. Then S_χ is locally isomorphic to $\mathbb{R}^3$ with coordinates (ξ_1, ξ_2, ξ_3). Let $f = (f_1, f_2, f_3) : X \to Y$ be a coordinates transformation on X to its copy Y with coordinates (y_1, y_2, y_3) which sends M_1 and M_2 to their copies defined by the same equations $\{y_2 = y_3 = 0\}$ and $\{y_1 = y_3 = 0\}$ respectively. Then the associated coordinates transformation from S_χ to its copy S_χ with coordinates (η_1, η_2, η_3) is given by*

$$\begin{aligned}
\eta_1 &= \frac{\partial f_1}{\partial x_1}(0)\xi_1, \\
\eta_2 &= \frac{\partial f_2}{\partial x_2}(0)\xi_2, \\
\eta_3 &= \frac{\partial f_3}{\partial x_3}(0)\xi_3 + \frac{\partial^2 f_3}{\partial x_1 \partial x_2}(0)\xi_1\xi_2.
\end{aligned}$$

Hence S_χ is not a vector bundle over $M = \{0\}$.

1.4 The local model

Throughout the paper we will often study the local model, i.e., the case in which $X = \mathbb{R}^n$ with coordinates $x = (x_1, \dots, x_n)$ and the submanifolds $M_j \in \chi$ $(j = 1, \dots, \ell)$ are defined by $\{x_i = 0,\, i \in I_j\}$ with $I_j \subseteq \{1, \dots, n\}$. In this setting we are able to study a more general situation we are going to explain below.

First of all, we may allow $I_j = I_k$ if $j \neq k$. The action associated with χ is now defined by $\mu_j(x, \lambda) = (\lambda^{a_{j1}} x_1, \dots, \lambda^{a_{jn}} x_n)$ with a_{ji} non-negative rational numbers, $a_{ji} \neq 0$ if $i \in I_j$, $a_{ji} = 0$ otherwise. We introduce such a situation from now.

Let $m + 1$ be the number of coordinates blocks, that is, coordinates (x) of $\mathbb{R}^n$ are divided into $(m + 1)$-coordinates blocks

$$(x^{(0)}, x^{(1)}, \dots, x^{(m)}).$$

Under these coordinates blocks, each M_j is assumed to be linearized, that is, for each M_j, there exists the subset $K_j \subset \{1, 2, \ldots, m\}$ such that

$$M_j = \{x^{(k)} = 0 \ (k \in K_j)\} \qquad (j = 1, 2, \ldots, \ell). \tag{1.11}$$

Let $A_\chi = (a_{ij})$ be an $\ell \times m$ matrix with $a_{ij} \in \mathbb{Q}_{\geq 0}$, and we take the positive rational number σ_A so that $\sigma_A a_{ij} \in \mathbb{Z}$ and all the $\sigma_A a_{ij}$ $(1 \leq i \leq \ell,\ 1 \leq j \leq m)$ have no common divisors.

Then we can define a more general normal deformation $\widetilde{X} = \mathbb{R}^n \times \mathbb{R}^\ell$ with the map $p : \widetilde{X} \to X$ defined by

$$p(x;\, t) = (x^{(0)},\, \varphi_1(t)^{\sigma_A} x^{(1)},\, \ldots,\, \varphi_m(t)^{\sigma_A} x^{(m)}) \tag{1.12}$$

with

$$\varphi_k(t) = \prod_{j=1}^{\ell} t_j^{a_{jk}} \qquad (k = 1, 2, \ldots, m). \tag{1.13}$$

Comparing with the matrix A_χ, when $t \in (\mathbb{R}^+)^\ell$ we have

$$(\log \varphi_1, \ldots, \log \varphi_m) = (\log t_1, \ldots, \log t_\ell) A_\chi.$$

In this setting, the action $\mu : \widetilde{X} \times (\mathbb{R}^+)^\ell \to \widetilde{X}$ is defined by the same equation as the one in (1.8), that is,

$$\mu((x;\, t), \lambda) = \left(x^{(0)},\, \varphi_1(\lambda) x^{(1)},\, \ldots,\, \varphi_m(\lambda) x^{(m)};\, \frac{t_1}{\lambda_1^{1/\sigma_A}},\, \ldots,\, \frac{t_\ell}{\lambda_\ell^{1/\sigma_A}}\right), \tag{1.14}$$

and $\mu_j : \widetilde{X} \times \mathbb{R}^+ \to \widetilde{X}$ for each j is obtained by

$$\mu_j((x;\, t), \lambda_j) = \left(x^{(0)},\, \lambda_j^{a_{j1}} x^{(1)},\, \ldots,\, \lambda_j^{a_{jm}} x^{(m)};\, t_1,\, \ldots,\, \frac{t_j}{1/\lambda_j^{\sigma_A}},\, \ldots,\, t_\ell\right) \tag{1.15}$$

for $\lambda_j \in \mathbb{R}^+$. Note that the actions μ and μ_j on X (which are denoted by the same symbols as those on $\widetilde{X}$) are given by

$$\begin{aligned} \mu(x, \lambda) &= \left(x^{(0)},\, \varphi_1(\lambda) x^{(1)},\, \ldots,\, \varphi_m(\lambda) x^{(m)}\right), \\ \mu_j(x, \lambda_j) &= \left(x^{(0)},\, \lambda_j^{a_{j1}} x^{(1)},\, \ldots,\, \lambda_j^{a_{jm}} x^{(m)}\right). \end{aligned} \tag{1.16}$$

Then we assume:

A2. Each M_j coincides with the set of fixed points of the action μ_j on X.

That is, we have

$$x \in M_j \iff \mu_j(x,\, \lambda_j) = x \text{ for any } \lambda_j \in \mathbb{R}^+. \tag{1.17}$$

In terms of A_χ, this condition is equivalent to saying that

$$a_{jk} \neq 0 \iff x^{(k)} = 0 \text{ on } M_j. \tag{1.18}$$

Example 1.16. *Let us see some examples in $\mathbb{R}^3$ with variables (x_1, x_2, x_3).*

- *Let $M_1 = \{x_1 = 0\}$, $M_2 = \{x_2 = x_3 = 0\}$ and consider the action on X by*

$$\mu(x,t) = (x_1 t_1, x_2 t_2, x_3 t_2).$$

Then we can take $x^{(1)} = x_1$, $x^{(2)} = (x_2, x_3)$ as the coordinates blocks and the associated matrix is

$$A_\chi = \begin{pmatrix} 1 & 0 \\ 0 & 1 \end{pmatrix}.$$

- *Let $M_1 = \{x_1 = x_2 = 0\}$, $M_2 = \{x_2 = 0\}$ and consider the action on X by*

$$\mu(x,t) = (x_1 t_1, x_2 t_1 t_2, x_3).$$

Then we can take $x^{(1)} = x_1$, $x^{(2)} = x_2$, $x^{(3)} = x_3$ as the coordinates blocks and the associated matrix is

$$A_\chi = \begin{pmatrix} 1 & 1 & 0 \\ 0 & 1 & 0 \end{pmatrix}.$$

- *Let $M_1 = M_2 = \{0\}$ and consider the action on X by*

$$\mu(x,t) = (x_1 t_1^3 t_2, x_2 t_1^3 t_2, x_3 t_1^2 t_2).$$

Then we can take $x^{(1)} = (x_1, x_2)$, $x^{(2)} = x_3$ as the coordinates blocks and the associated matrix is

$$A_\chi = \begin{pmatrix} 3 & 2 \\ 1 & 1 \end{pmatrix}.$$

- *Let $M_1 = \{x_1 = x_2 = 0\}$, $M_2 = \{x_2 = x_3 = 0\}$ and consider the action on X by*

$$\mu(x,t) = (x_1 t_1, x_2 t_1 t_2, x_3 t_2).$$

Then we can take $x^{(1)} = x_1$, $x^{(2)} = x_2$, $x^{(3)} = x_3$ as the coordinates blocks and the associated matrix is

$$A_\chi = \begin{pmatrix} 1 & 1 & 0 \\ 0 & 1 & 1 \end{pmatrix}.$$

Finally, in a local model, we use the same terminologies as in Definition 1.3 and Lemma 1.5. That is,

Definition 1.17. *We say that μ_χ has maximal rank if* $\operatorname{Rank} A_\chi = \min\{\ell, m\}$ *holds for the associated matrix A_χ of μ_χ. It is said to be non-degenerate if* $\operatorname{Rank} A_\chi = \ell$ *and transitive if* $\operatorname{Rank} A_\chi = m$.

2 Multi-cones for a general case

In this section we find a family of multi-cones associated with a multi-normal deformation. We mainly work in the local model introduced in Subsection 1.4.

2.1 Fixed points of S_χ

As we have seen in the previous section, if χ is of transitive type, S_χ is the Cartesian product of vector bundles over M. Hence we can consider the subbundle consisting of points $\xi = \xi^{(1)} \times_M \cdots \times_M \xi^{(m)}$ where some of $\xi^{(k)}$'s belong to the zero section of the corresponding vector bundles. Here we introduce its counterpart for a general case.

For $a = (a_1, \ldots, a_\ell) \in \mathbb{R}^\ell$, we define the action $\mu^a : S_\chi \times \mathbb{R}^+ \to S_\chi$ by

$$\mu^a(p, t) := \mu(p, t^{a_1}, \ldots, t^{a_\ell}) \qquad (t \in \mathbb{R}^+).$$

Definition 2.1. *We say that a point $p \in S_\chi$ is a fixed point (with respect to the action μ) if there exists $a = (a_1, \ldots, a_\ell) \in \mathbb{R}^\ell$ for which μ^a is not the identity action on S_χ and p is a fixed point of μ^a, that is, $\mu^a(\cdot, t) \neq \mathrm{id}_{S_\chi}$ for some $t > 0$ and $\mu^a(p, t) = p$ for any $t > 0$.*

We give some examples.

Example 2.2. *Let χ be a linearizable family of closed submanifolds. We assume χ to be of transitive type. Then S_χ has locally a system of coordinates blocks $(x^{(0)}; \xi^{(1)}, \ldots, \xi^{(m)})$ as described in the previous section, where m is the rank of the family χ. In this coordinates, the set of fixed points is given by $\{|\xi^{(1)}||\xi^{(2)}| \cdots |\xi^{(m)}| = 0\}$. Hence a point is located outside the set of fixed points if and only if all the $\xi^{(k)}$'s are different from the origin.*

Example 2.3. *Let us consider the situation observed in Example 1.15. In this case, the set of fixed points is given by*

$$\{\xi_1 = \xi_2 = 0 \ \textit{or}\ \xi_2 = \xi_3 = 0 \ \textit{or}\ \xi_1 = \xi_3 = 0\}.$$

Hence a point is outside the set of fixed points if and only if at least two ξ_i's are non-zero.

Let us consider a local model described in Subsection 1.4, where $m + 1$ denotes the number of coordinates blocks and ℓ designates the number of manifolds in χ. Set $L := \mathrm{Rank}\, A_\chi$. Note that $L \le \min\{\ell, m\}$ holds.

Lemma 2.4. *Let $p = (x^{(0)}, \xi^{(1)}, \ldots, \xi^{(m)}) \in S_\chi$ be a point outside the set of fixed points. Let $1 \le j_1 < \cdots < j_L \le \ell$ and assume that the j_i-th rows $(1 \le i \le L)$ are linearly independent. Then there exist $1 \le k_1 < \cdots < k_L \le m$ which satisfy the following conditions.*

1. *$\xi^{(k_i)} \neq 0$ $(1 \leq i \leq L)$.*
2. *The $L \times L$ submatrix made from the j_i-th rows and the k_i-th columns $(1 \leq i \leq L)$ in $A_{\mathcal{X}}$ is invertible.*

Proof. By taking j_i-th rows $(1 \leq i \leq L)$ in $A_{\mathcal{X}}$, we may assume that the size of $A_{\mathcal{X}}$ is $L \times m$. By a permutation of columns, we may further assume $\xi^{(k)} = 0$ $(k > m')$ for some m'. Let $\widetilde{A}$ be the $L \times m'$ submatrix of $A_{\mathcal{X}}$ consisting of the first m'-columns. If any choice of L-columns in $\widetilde{A}$ gives linearly dependent column vectors, then the rank of $\widetilde{A}$ is less than L, and hence, we can find a non-zero row vector $a = (a_1, \ldots, a_L) \in \mathbb{R}^L$ with $a\widetilde{A} = 0$. Then the associated action μ^a is different from $\mathrm{id}_{S_{\mathcal{X}}}$ and $\mu^a(p,t) = p$ holds for any $t > 0$. This contradicts the fact that p is not a fixed point. Hence we can find L-columns which satisfy the claim of the lemma. □

Corollary 2.5. *Let $p = (x^{(0)}, \xi^{(1)}, \ldots, \xi^{(m)}) \in S_{\mathcal{X}}$ be a point outside the set of fixed points. Then, by an appropriate permutation of coordinates blocks and that of the parameters in the action μ, we have the followings:*

1. *The first L-directions of p are different from the origin, i.e., $\xi^{(k)} \neq 0$ $(1 \leq k \leq L)$.*
2. *The $L \times L$ submatrix made from the first L-columns and the first L-rows in $A_{\mathcal{X}}$ is invertible.*

Example 2.6. *Let $M_1 = \{x_1 = x_2 = 0\}$, $M_2 = \{x_2 = x_3 = 0\}$ and consider the action on X by*

$$\mu(x,t) = (x_1 t_1, x_2 t_1 t_2, x_3 t_2).$$

Then we can take $x^{(1)} = x_1$, $x^{(2)} = x_2$, $x^{(3)} = x_3$ as the coordinates blocks and the associated matrix is

$$A_{\mathcal{X}} = \begin{pmatrix} 1 & 1 & 0 \\ 0 & 1 & 1 \end{pmatrix}.$$

We have seen that $p = (\xi^{(1)}, \xi^{(2)}, \xi^{(3)}) \in S_{\mathcal{X}}$ is outside the set of fixed points if two of the three coordinates are different from 0.

- *If $\xi^{(1)}, \xi^{(2)} \neq 0$, then the invertible 2×2 matrix is*

$$\begin{pmatrix} 1 & 1 \\ 0 & 1 \end{pmatrix}.$$

- *If $\xi^{(1)}, \xi^{(3)} \neq 0$, then the invertible 2×2 matrix is*

$$\begin{pmatrix} 1 & 0 \\ 0 & 1 \end{pmatrix}.$$

- *If $\xi^{(2)}, \xi^{(3)} \neq 0$, then the invertible 2×2 matrix is*

$$\begin{pmatrix} 1 & 0 \\ 1 & 1 \end{pmatrix}.$$

2.2 Semigroups of generators

From now on, we consider the problem under the local model described in Section 1.4. Let $m+1$ be the number of coordinates blocks and ℓ the number of submanifolds in χ. Set $L := \operatorname{Rank} A_\chi$ $(1 \leq L \leq \min\{\ell, m\})$. By an appropriate permutation of coordinates blocks and that of parameters of the action μ, we assume:

$$\begin{aligned}&\text{The first } L \times L \text{ submatrix in } A_\chi = (a_{jk}) \text{ is invertible,}\\ &\text{that is, } \det\,(a_{jk})_{1\leq j,k\leq L} \neq 0.\end{aligned} \tag{2.1}$$

Take a point

$$p = (x^{(0)};\, \xi) = (x^{(0)};\, \xi^{(1)},\, \ldots,\, \xi^{(m)}) \in S_\chi. \tag{2.2}$$

Then we also assume:

$$\text{If the } k\text{-th column of } A_\chi \text{ is a zero vector, the corresponding } \xi^{(k)} \text{ is zero.} \tag{2.3}$$

By the assumption (2.1), we may assume that p is normalized as

$$|\xi^{(k)}| = 1 \text{ for } 1 \leq k \leq L \text{ with } \xi^{(k)} \neq 0. \tag{2.4}$$

Note that we cannot assume this normalization for indices $k > L$.

Recall that the action on X is given by

$$(x^{(0)},\, \varphi_1(\lambda)x^{(1)}, \ldots, \varphi_m(\lambda)x^{(m)})$$

for $\lambda = (\lambda_1,\, \ldots,\, \lambda_\ell)$, where m-functions $\varphi_1,\, \ldots,\, \varphi_m$ are monomials of λ defined by

$$\varphi_k(\lambda) = \lambda_1^{a_{1k}} \cdots \lambda_\ell^{a_{\ell k}} \qquad (k = 1, \ldots, m).$$

Here a_{jk}'s are entries of $A_\chi = (a_{jk})$ with $a_{jk} \in \mathbb{Q}_{\geq 0}$.

Let τ denote m-variables $(\tau_1,\, \ldots,\, \tau_m)$, and we denote by τ' the first L-variables of τ and by τ'' the rest, that is, we have $\tau = (\tau',\, \tau'')$ with

$$\tau' = (\tau_1, \ldots, \tau_L) \;\text{ and }\; \tau'' = (\tau_{L+1}, \ldots, \tau_m). \tag{2.5}$$

In the same way, we denote by λ' the first L-variables of λ and by λ'' the rest, that is, $\lambda = (\lambda',\, \lambda'')$ with

$$\lambda' = (\lambda_1, \ldots, \lambda_L) \;\text{ and }\; \lambda'' = (\lambda_{L+1}, \ldots, \lambda_\ell). \tag{2.6}$$

Consider the L-equations of λ'

$$\tau' = \varphi'(\lambda',\, \lambda''), \tag{2.7}$$

where φ' denotes $(\varphi_1,\, \ldots,\, \varphi_L)$. Then, by the assumption, the system of the equations can be solved for the variables $\lambda' = (\lambda_1, \ldots, \lambda_L)$. That is, we can find

$$\lambda' = \varphi^{-1}(\tau',\, \lambda'') = (\varphi_1^{-1}(\tau',\, \lambda''),\, \ldots,\, \varphi_L^{-1}(\tau',\, \lambda'')) \tag{2.8}$$

which satisfies

$$\tau_k = \varphi_k(\varphi^{-1}(\tau', \lambda''), \lambda'') \qquad (k = 1, \dots, L). \tag{2.9}$$

Let $\mathcal{H}_\tau$ (resp. $\mathcal{H}_{\tau,\lambda}$) be the set of non-zero monomials of rational powers of τ (resp. τ and λ) with coefficients 1, that is, each element has a form

$$\begin{aligned} &\tau_1^{\alpha_1} \cdots \tau_m^{\alpha_m} \quad (\alpha_1, \dots, \alpha_m \in \mathbb{Q}) \\ &\big(\text{resp. } \tau_1^{\alpha_1} \cdots \tau_m^{\alpha_m} \lambda_1^{\beta_1} \cdots \lambda_\ell^{\beta_\ell} \quad (\alpha_1, \dots, \alpha_m, \beta_1, \dots, \beta_\ell \in \mathbb{Q})\big). \end{aligned} \tag{2.10}$$

For $f \in \mathcal{H}_{\tau,\lambda}$ given in (2.10), we denote by $\nu_k(f)$ (resp. $\nu_j^\lambda(f)$) the exponent of the variable τ_k (resp. λ_j) in f, i.e.,

$$\nu_k(f) = \alpha_k \quad \text{and} \quad \nu_j^\lambda(f) = \beta_j. \tag{2.11}$$

Then we regard $\mathcal{H}_{\tau,\lambda} \times \mathbb{R}_{\geq 0}$ to be a semigroup with its multiplication defined by, for pairs (f, v) and $(g, w) \in \mathcal{H}_{\tau,\lambda} \times \mathbb{R}_{\geq 0}$

$$(f, v)(g, w) = (fg, vw). \tag{2.12}$$

Let $\mathcal{A}$ and $\mathcal{B}$ be semigroups in $\mathcal{H}_{\tau,\lambda} \times \mathbb{R}_{\geq 0}$. Then we define the semigroup $\mathcal{A} \vee \mathcal{B}$ as the minimal semigroup in $\mathcal{H}_{\tau,\lambda} \times \mathbb{R}_{\geq 0}$ containing $\mathcal{A}$ and $\mathcal{B}$, that is,

$$\mathcal{A} \vee \mathcal{B} = \{f, g, fg;\ f \in \mathcal{A},\ g \in \mathcal{B}\}. \tag{2.13}$$

We also define the semigroup

$$\mathcal{A} * \mathcal{B} := \{fg;\ f \in \mathcal{A},\ g \in \mathcal{B}\}. \tag{2.14}$$

For example, $\{(1, 0)\}$ is a semigroup in $\mathcal{H}_{\tau,\lambda} \times \mathbb{R}_{\geq 0}$ which consists of only one element $(1, 0)$. Then

$$\mathcal{B} * \{(1, 0)\} = \{(f, 0);\ (f, v) \in \mathcal{B}\}.$$

We can also regard $\mathcal{H}_\tau \times \mathbb{R}_{\geq 0} \subset \mathcal{H}_{\tau,\lambda} \times \mathbb{R}_{\geq 0}$ to be a semigroup and define the corresponding operations on $\mathcal{H}_\tau \times \mathbb{R}_{\geq 0}$ induced from those on $\mathcal{H}_{\tau,\lambda} \times \mathbb{R}_{\geq 0}$.

Set, for $p = (x^{(0)}; \xi) = (x^{(0)}; \xi^{(1)}, \dots, \xi^{(m)}) \in S_\chi$ given in (2.2),

$$J_Z(\xi) := \{k \in \{1, 2, \dots, m\};\ \xi^{(k)} = 0\}, \tag{2.15}$$

and assume

$$\begin{aligned} &J_Z(\xi) \cap \{1, 2, \dots, L\} = \{k_1, k_2, \dots, k_q\} \\ &(1 \leq k_1 < k_2 < \cdots < k_q \leq L), \end{aligned} \tag{2.16}$$

where $q := \#\left(J_Z(\xi) \cap \{1, 2, \dots, L\}\right) \geq 0$. We define, for a subset A in $\mathcal{H}_{\tau,\lambda} \times \mathbb{R}_{\geq 0}$,

$$A_{\geq} := \left\{(f,\, v) \in A;\, \nu_{k_i}(f) \geq 0 \ (1 \leq i \leq q)\right\} \bigcap (\mathcal{H}_\tau \times \mathbb{R}_{\geq 0}), \tag{2.17}$$

$$A_{=} := \left\{(f,\, v) \in A;\, \nu_{k_i}(f) = 0 \ (1 \leq i \leq q)\right\} \bigcap (\mathcal{H}_\tau \times \mathbb{R}_{\geq 0}), \tag{2.18}$$

and

$$A_{>} := A_{\geq} \setminus A_{=} \ \subset (\mathcal{H}_\tau \times \mathbb{R}_{\geq 0}). \tag{2.19}$$

Note that, if $\mathcal{B}$ is a semigroup in $\mathcal{H}_{\tau,\lambda} \times \mathbb{R}_{\geq 0}$, then $\mathcal{B}_{\geq}$, $\mathcal{B}_{=}$ and $\mathcal{B}_{>}$ are semigroups in $\mathcal{H}_\tau \times \mathbb{R}_{\geq 0}$. For a subset $A \subset \mathcal{H}_{\tau,\lambda} \times \mathbb{R}_{\geq 0}$, we introduce its fraction $\mathrm{Q}(A) \subset \mathcal{H}_{\tau,\lambda} \times \mathbb{R}_{\geq 0}$ by

$$A \bigcup \{(f^{-1},\, v^{-1})\}_{\{(f,\, v) \in A;\, v \neq 0\}}\,. \tag{2.20}$$

Definition 2.7. *Let $\mathcal{A}$ and $\mathcal{B}$ be semigroups in $\mathcal{H}_\tau \times \mathbb{R}_{\geq 0}$. We say that $\mathcal{A}$ and $\mathcal{B}$ are equivalent if and only if there exist natural numbers N_1 and N_2 with*

$$\langle \mathcal{A} \rangle_{N_1} \ \subset \ \langle \mathcal{B} \rangle_{N_2} \ \subset \ \langle \mathcal{A} \rangle. \tag{2.21}$$

Here $\langle \mathcal{A} \rangle_N$ denotes the semigroup

$$\langle \mathcal{A} \rangle_N \ = \left\{(f,\, v)^N;\, (f,\, v) \in \mathcal{A}\right\} \subset \mathcal{A}. \tag{2.22}$$

Remark 2.8. *If $\langle \mathcal{G} \rangle_N = \mathcal{A}$ $(N \in \mathbb{N})$ holds for semigroups $\mathcal{A} \subset \mathcal{G}$, we sometimes say that the radical of $\mathcal{A}$ is $\mathcal{G}$.*

Now recall that $\varphi^{-1}(\tau', \lambda'')$ was defined in (2.8). Then we define

$$\psi_k(\tau,\, \lambda'') := \frac{\tau_k}{\varphi_k(\varphi^{-1}(\tau',\, \lambda''),\, \lambda'')} \qquad (L < k \leq m). \tag{2.23}$$

Lemma 2.9. *Let $L < k \leq m$. Then $\varphi_k(\varphi^{-1}(\tau',\, \lambda''),\, \lambda'')$ depends only on the variables τ'. In particular, $\psi_k(\tau,\, \lambda'')$ does not depend on the variables λ''.*

Proof. The k-th column of A_χ is a linear combination of the first L-columns in A_χ, and thus, there exist $\alpha_{kj} \in \mathbb{Q}$ satisfying

$$\varphi_k = \prod_{1 \leq j \leq L} \varphi_j^{\alpha_{kj}} \qquad (L < k \leq m).$$

Hence we have

$$\begin{aligned}\varphi_k(\varphi^{-1}(\tau', \lambda''), \lambda'') &= \prod_{1 \leq j \leq L} \varphi_j(\varphi^{-1}(\tau', \lambda''), \lambda'')^{\alpha_{kj}} \\ &= \prod_{1 \leq j \leq L} \tau_j^{\alpha_{kj}}.\end{aligned}$$

□

By the lemma, $\psi_k(\tau, \lambda'')$ is denoted by $\psi_k(\tau)$ $(L < k \leq m)$ in what follows. Now we introduce the finite subset G in $\mathcal{H}_{\tau,\lambda} \times \mathbb{R}_{\geq 0}$ defined by

$$G := Q\left(\left\{(\varphi_j^{-1}, 0)\right\}_{1\leq j\leq L} \bigcup \left\{\left(\psi_k, |\xi^{(k)}|\right)\right\}_{L<k\leq m} \bigcup \{(\lambda_j, 0)\}_{L<j\leq \ell}\right). \quad (2.24)$$

Remark 2.10. *In the definition of G, we use the fact that the point ξ is normalized, that is, $|\xi^{(k)}| = 1$ for $1 \leq k \leq L$ with $\xi^{(k)} \neq 0$. When ξ is not normalized, we define G by*

$$G := Q\left(\left\{(\varphi_j^{-1}, 0)\right\}_{1\leq j\leq L} \bigcup \left\{\, (\psi_k, n_k(\xi))\,\right\}_{L<k\leq m} \bigcup \{(\lambda_j, 0)\}_{L<j\leq \ell}\right). \quad (2.25)$$

Here we set

$$n_k(\xi) := \psi_k\left(|\xi^{(*)}|_{+,1}, |\xi^{(*)}|_{+,2}, \ldots, |\xi^{(*)}|_{+,m}\right), \quad (2.26)$$

where $|\xi^{(*)}|_{+,k} = |\xi^{(k)}|$ if $k \notin J_Z(\xi) \cap \{1, \ldots, L\}$ and $|\xi^{(*)}|_{+,k} = 1$ otherwise.

Remark 2.11. *If the associated action μ has maximal rank, the definition of G becomes much simpler. In fact, when the action μ is non-degenerate, we have*

$$G = Q\left(\left\{(\varphi_j^{-1}(\tau'), 0)\right\}_{1\leq j\leq \ell} \bigcup \left\{\, (\psi_k(\tau), n_k(\xi))\,\right\}_{\ell<k\leq m}\right). \quad (2.27)$$

Note that, in this case, G is contained in $\mathcal{H}_\tau \times \mathbb{R}_{\geq 0}$, that is, the variables λ do not appear. On the other hand, when the action μ is transitive, we have

$$G = Q\left(\left\{(\varphi_j^{-1}(\tau, \lambda''), 0)\right\}_{1\leq j\leq m} \bigcup \{(\lambda_j, 0)\}_{m<j\leq \ell}\right). \quad (2.28)$$

In this case, the function ψ_k does not appear.

We then define the semigroup in $\mathcal{H}_\tau \times \mathbb{R}_{\geq 0}$ by

$$\mathcal{G} := [G]_{=} \bigvee \left([G]_{>} * \{(1, 0)\}\right). \quad (2.29)$$

Here $[B]$ denotes the semigroup in $\mathcal{H}_{\tau,\lambda} \times \mathbb{R}_{\geq 0}$ generated by a subset B, that is, if B consists of (f_s, v_s) $(s \in \Lambda)$ for a finite subset Λ, then

$$[B] := \left\{\prod_{s\in\Lambda} (f_s, v_s)^{\alpha_s} \in \mathcal{H}_{\tau,\lambda} \times \mathbb{R}_{\geq 0};\ \alpha_s \in \mathbb{Z}_{\geq 0}\ \ (s \in \Lambda)\right\}. \quad (2.30)$$

Note that $[B]$ contains the unit $(1, 1)$ because we allow all the α_s to be zero.

Example 2.12. *Let $M_1 = \{x_1 = x_2 = 0\}$, $M_2 = \{x_2 = x_3 = 0\}$ and consider the action on X by*

$$\mu(x, t) = (x_1t_1, x_2t_1t_2, x_3t_2).$$

Then we can take $x^{(1)} = x_1$, $x^{(2)} = x_2$, $x^{(3)} = x_3$ as the coordinates blocks and the associated matrix is

$$A_X = \begin{pmatrix} 1 & 1 & 0 \\ 0 & 1 & 1 \end{pmatrix}.$$

We have $\varphi_1(\lambda) = \lambda_1$, $\varphi_2(\lambda) = \lambda_1\lambda_2$, $\varphi_3(\lambda) = \lambda_2$. *The invertible* 2×2 *matrix is*

$$A'_\chi = \begin{pmatrix} 1 & 1 \\ 0 & 1 \end{pmatrix}, \qquad A'^{\,-1}_\chi = \begin{pmatrix} 1 & -1 \\ 0 & 1 \end{pmatrix}.$$

We have $\varphi_1^{-1}(\tau) = \tau_1$, $\varphi_2^{-1}(\tau) = \dfrac{\tau_2}{\tau_1}$, $\psi_3(\tau) = \dfrac{\tau_3\tau_1}{\tau_2}$. *Then*

$$G = Q\left(\left\{(\tau_1, 0), \left(\frac{\tau_2}{\tau_1}, 0\right), \left(\frac{\tau_1\tau_3}{\tau_2}, |\xi_3|\right)\right\}\right),$$

that is, we have

$$G = \left\{(\tau_1, 0), \left(\frac{\tau_2}{\tau_1}, 0\right), \left(\frac{\tau_1\tau_3}{\tau_2}, |\xi_3|\right), \left(\frac{\tau_2}{\tau_1\tau_3}, \frac{1}{|\xi_3|}\right)\right\} \qquad \text{if } \xi_3 \neq 0$$

and

$$G = \left\{(\tau_1, 0), \left(\frac{\tau_2}{\tau_1}, 0\right), \left(\frac{\tau_1\tau_3}{\tau_2}, 0\right)\right\} \qquad \text{if } \xi_3 = 0.$$

Now we are going to define $\mathcal{G}$. *Remark that, by definition, all the couples* $(\tau_i, 0)$, $i = 1, 2, 3$ *and* $\left(\dfrac{\tau_2}{\tau_3}, 0\right)$ *belong to* $[G]$. *Let* $p = (\xi_1, \xi_2, \xi_3) \in S_\chi$.

- *If* $\xi_1, \xi_2, \xi_3 \neq 0$, *then* $\mathcal{G}$ *is generated by the products of the elements of* G.
- *If* $\xi_1 = 0$, $\xi_2, \xi_3 \neq 0$, *then* $\mathcal{G}$ *is generated by the products of the family*

$$\left\{(\tau_1, 0), (\tau_2, 0), (\tau_3, 0), \left(\frac{\tau_2}{\tau_3}, 0\right), \left(\frac{\tau_1\tau_3}{\tau_2}, 0\right), (1, 1)\right\}.$$

- *If* $\xi_2 = 0$, $\xi_1, \xi_3 \neq 0$, *then* $\mathcal{G}$ *is generated by the products of the family*

$$\left\{(\tau_1, 0), (\tau_2, 0), (\tau_3, 0), \left(\frac{\tau_2}{\tau_1}, 0\right), \left(\frac{\tau_2}{\tau_1\tau_3}, 0\right), (1, 1)\right\}.$$

- *If* $\xi_3 = 0$, $\xi_1, \xi_2 \neq 0$, *then* $\mathcal{G}$ *is generated by the products of the family*

$$\left\{(\tau_1, 0), (\tau_2, 0), (\tau_3, 0), \left(\frac{\tau_2}{\tau_1}, 0\right), \left(\frac{\tau_1\tau_3}{\tau_2}, 0\right), (1, 1)\right\}.$$

- *If* $\xi_1, \xi_2 = 0$, $\xi_3 \neq 0$, *then* $\mathcal{G}$ *is generated by the products of the family*

$$\left\{(\tau_1, 0), (\tau_2, 0), (\tau_3, 0), \left(\frac{\tau_2}{\tau_3}, 0\right), (1, 1)\right\}.$$

- *If* $\xi_2, \xi_3 = 0$, $\xi_1 \neq 0$, *then* $\mathcal{G}$ *is generated by the products of the family*

$$\left\{(\tau_1, 0), (\tau_2, 0), (\tau_3, 0), \left(\frac{\tau_2}{\tau_1}, 0\right), (1, 1)\right\}.$$

- *If* $\xi_1, \xi_3 = 0$, $\xi_2 \neq 0$, *then* $\mathcal{G}$ *is generated by the products of the family*

$$\left\{(\tau_1, 0), (\tau_2, 0), (\tau_3, 0), \left(\frac{\tau_1\tau_3}{\tau_2}, 0\right), (1, 1)\right\}.$$

- *If* $\xi_1, \xi_2, \xi_3 = 0$, *then* $\mathcal{G}$ *is generated by the products of the family*

$$\{(\tau_1, 0), (\tau_2, 0), (\tau_3, 0), (1, 1)\}.$$

The semigroup $\mathcal{G}$ seems to depend on the choice of linearly independent L-columns in A_χ. However, by the following lemma, we see that it is essentially independent of such a choice for our purpose. Define the finite subset $\widehat{G} \subset \mathcal{H}_{\tau,\lambda} \times \mathbb{R}_{\geq 0}$ by

$$Q\left(\left\{\left(\frac{\tau_1}{\varphi_1(\lambda)}, |\xi^{(1)}|\right), \cdots, \left(\frac{\tau_m}{\varphi_m(\lambda)}, |\xi^{(m)}|\right), (\lambda_1, 0), \cdots, (\lambda_\ell, 0)\right\}\right), \tag{2.31}$$

and set

$$\widehat{\mathcal{G}} := (\mathcal{H}_\tau \times \mathbb{R}_{\geq 0}) \bigcap [\widehat{G}]. \tag{2.32}$$

Proposition 2.13. *The semigroups* $\mathcal{G}$ *and* $\widehat{\mathcal{G}}$ *are equivalent. To be more precise, we have*

$$\langle \mathcal{G} \rangle_N \subset \widehat{\mathcal{G}} \subset \mathcal{G} \tag{2.33}$$

for some $N \in \mathbb{N}$.

In this proof, we adopt (2.25) as the definition of G, that is, the normalization of ξ is not assumed. Set

$$g_k(\tau, \lambda) := \frac{\tau_k}{\varphi_k(\lambda)} \quad (k = 1, \ldots, m).$$

Then the proposition is a consequence of the following 3 lemmas.

Lemma 2.14. *We have*

$$(\mathcal{H}_\tau \times \mathbb{R}_{\geq 0}) \bigcap [\widehat{G}] = \widehat{\mathcal{G}} \subset \mathcal{G}.$$

Proof. Take $(f, v) \in (\mathcal{H}_\tau \times \mathbb{R}_{\geq 0}) \bigcap [\widehat{G}]$. Then, by definition, there exist $\alpha_k, \beta_j \in \mathbb{Z}$ such that

$$(f(\tau), v) = \prod_{1 \leq k \leq m} (g_k, |\xi^{(k)}|)^{\alpha_k} \prod_{1 \leq j \leq \ell} (\lambda_j, 0)^{\beta_j}.$$

Here $\alpha_k \geq 0$ if $k \in J_Z(\xi)$ and $\beta_j \geq 0$ $(1 \leq j \leq \ell)$. Note that it follows from construction of $\widehat{G}$ that $\nu_k(f) \geq 0$ for any $k \in J_Z(\xi)$. Since $f(\tau)$ does not depend on λ, by putting $\lambda' = \varphi^{-1}(\tau', \lambda'')$ into the above equation, we obtain

$$f = \prod_{1 \leq j \leq L} \varphi_j^{-1}(\tau', \lambda'')^{\beta_j} \prod_{L < k \leq m} \psi_k(\tau)^{\alpha_k} \prod_{L < j \leq \ell} \lambda_j^{\beta_j}.$$

Set

$$(f, v') = \prod_{1 \leq j \leq L} (\varphi_j^{-1}, 0)^{\beta_j} \prod_{L < k \leq m} (\psi_k, n_k(\xi))^{\alpha_k} \prod_{L < j \leq \ell} (\lambda_j, 0)^{\beta_j}.$$

Note $f \in \mathcal{H}_\tau$. Then it follows from the definition of $\mathcal{G}$ that we have $(f, v') \in \mathcal{G}$ if $\nu_k(f) = 0$ for all $k \in J_Z(\xi) \cap \{1, \ldots, L\}$ and $(f, 0) = (f, v') * (1, 0) \in \mathcal{G}$ if $\nu_k(f) > 0$

for some $k \in J_Z(\xi) \cap \{1, \ldots, L\}$. Hence, to show $(f, v) \in \mathcal{G}$, it suffices to prove $v = v'$ for the former case and $v = 0$ for the latter case.

Assume $\nu_k(f) > 0$ for some $k \in J_Z(\xi) \cap \{1, \ldots, L\}$. Since only g_k contains the variable τ_k, the corresponding α_k must be greater than 0, and thus, v becomes zero. This implies $(f, v) \in [G]_> * [(1,0)]$.

Assume $\nu_k(f) = 0$ for any $k \in J_Z(\xi) \cap \{1, \ldots, L\}$. Since $\alpha_k = 0$ for any $k \in J_Z(\xi) \cap \{1, \ldots, L\}$, if $v = 0$, then some β_j or some α_k with $k \in J_Z(\xi) \cap \{L+1, \ldots, m\}$ is non-zero. Hence, we have $v' = 0$ and we conclude $(f, v) \in [G]_=$ when $v = 0$.

When $v \neq 0$, we have $\beta_j = 0$ for $1 \leq j \leq \ell$ and $\alpha_k = 0$ for $k \in J_Z(\xi)$, which implies $v' \neq 0$. Then $(f, v) \in [G]_=$ immediately follows from the following elementary lemma. Hence we obtained the desired inclusion. □

Lemma 2.15. *Let $(h, w) \in \mathcal{G}$ with $w \neq 0$. If h can be written in the form $\prod_{1 \leq k \leq m} \tau_k^{\delta_k}$ for some $\delta_k \in \mathbb{Q}$ $(1 \leq k \leq m)$, then we have $w = \prod_{1 \leq k \leq m} |\xi^{(k)}|^{\delta_k}$. The same fact holds for an element $(h, w) \in \widehat{\mathcal{G}}$ with $w \neq 0$.*

Proof. Since $w \neq 0$, we have $(h, w) \in [G]_=$ and it is a product of integer powers of $(\psi_k, n_k(\xi))$, i.e., there exist $\alpha_k \in \mathbb{Z}$ such that

$$(h, w) = \prod_{L < k \leq m} (\psi_k, n_k(\xi))^{\alpha_k}. \tag{2.34}$$

We temporarily regard $|\xi^{(1)}|, \ldots, |\xi^{(m)}|$ as variables and, for $1 \leq i \leq m$, we define ν_i^ξ in the similar way as that for ν_i, i.e., $\nu_i^\xi(g)$ denotes the exponent of $|\xi^{(i)}|$ in a polynomial g of variables $|\xi^{(1)}|, \ldots, |\xi^{(m)}|$. Then, by the definition of $n_k(\xi)$, we have, for $1 \leq k \leq m$,

$$\nu_i(\psi_k) = \nu_i^\xi(n_k(\xi)) \qquad (i \notin J_Z(\xi) \cap \{1, \ldots, L\})$$

and

$$\nu_i^\xi(n_k(\xi)) = 0 \qquad (i \in J_Z(\xi) \cap \{1, \ldots, L\}).$$

Then, by (2.34), we get

$$\nu_i(h) = \nu_i^\xi(w(\xi)) \qquad (i \notin (J_Z(\xi) \cap \{1, \ldots, L\}).$$

Since $(h, w) \in [G]_=$ implies $\nu_i(h) = 0$ for $i \in J_Z(\xi) \cap \{1, \ldots, L\}$, we finally obtain

$$\nu_i(h) = \nu_i^\xi(w(\xi)) \qquad (1 \leq i \leq m),$$

from which we get the first claim.

Since $(h, w) \in \widehat{\mathcal{G}}$ with $w \neq 0$ implies that h is a product of integer powers of $\tau_k/\varphi_k(\lambda)$'s $(1 \leq k \leq m)$, i.e., no factors $(\lambda_j, 0)$ appear, we can easily conclude the second claim. □

Now we will prove the converse inclusion.

Lemma 2.16. *We have*

$$\langle\mathcal{G}\rangle_N \subset \widehat{\mathcal{G}} = (\mathcal{H}_\tau \times \mathbb{R}_{\geq 0}) \bigcap [\widehat{G}]$$

for some $N \in \mathbb{N}$.

Proof. It follows from the relations

$$\varphi_j^{-1}(\varphi'(\lambda', \lambda''), \lambda'') = \lambda_j \qquad (1 \leq j \leq L)$$

that $\dfrac{\varphi_j^{-1}(\tau', \lambda'')}{\lambda_j}$ $(1 \leq j \leq L)$ is expressed by a product of rational powers of g_1, $\ldots$, g_L, that is, there exist $\gamma_{jk} \in \mathbb{Q}$ such that

$$\frac{\varphi_j^{-1}(\tau', \lambda'')}{\lambda_j} = \prod_{1 \leq k \leq L} g_k^{\gamma_{jk}} \qquad (1 \leq j \leq L), \tag{2.35}$$

equivalently we have

$$\varphi_j^{-1}(\tau', \lambda'') = \left(\prod_{1 \leq k \leq L} g_k^{\gamma_{jk}}\right) \lambda_j \qquad (1 \leq j \leq L). \tag{2.36}$$

By using (2.35), we see that ψ_k is also expressed by a product of rational powers of $g_1, \ldots, g_m$, that is, there exist $\gamma'_{jk} \in \mathbb{Q}$ such that

$$\psi_j = \prod_{1 \leq k \leq m} g_k^{\gamma'_{jk}} \qquad (L < j \leq m). \tag{2.37}$$

Let $(f, v) \in \mathcal{G}$. By definition, there exist α_k and $\beta_j \in \mathbb{Z}$ such that

$$(f, v) = \prod_{1 \leq k \leq L} (\varphi_k^{-1}, 0)^{\alpha_k} \prod_{L < k \leq m} (\psi_k, n_k(\xi))^{\alpha_k} \prod_{L < j \leq \ell} (\lambda_j, 0)^{\beta_j}$$

holds if $\nu_k(f) = 0$ for any $k \in J_Z(\xi) \cap \{1, 2, \ldots, L\}$, and

$$(f, v) = \prod_{1 \leq k \leq L} (\varphi_k^{-1}, 0)^{\alpha_k} \prod_{L < k \leq m} (\psi_k, n_k(\xi))^{\alpha_k} \prod_{L < j \leq \ell} (\lambda_j, 0)^{\beta_j} * (1, 0)$$

holds if $\nu_k(f) > 0$ for some $k \in J_Z(\xi) \cap \{1, 2, \ldots, L\}$. Here $\beta_j \geq 0$ for any j and $\alpha_k \geq 0$ if $1 \leq k \leq L$ or if $k > L$ with $k \in J_Z(\xi)$.

Then, by putting (2.36) and (2.37) into the above f, we can find $\delta_k \in \mathbb{Q}$ satisfying

$$f = \prod_{1 \leq k \leq m} g_k^{\delta_k} \prod_{1 \leq k \leq L} \lambda_k^{\alpha_k} \prod_{L < j \leq \ell} \lambda_j^{\beta_j}.$$

Since only g_k contains the variable τ_k, we know $\delta_k = \nu_k(f)$ $(1 \le k \le m)$. In particular, we have $\delta_k \ge 0$ for $k \in J_Z(\xi)$. Set

$$(f,\, v') = \prod_{1\le k\le m} (g_k,\, |\xi^{(k)}|)^{\delta_k} \prod_{1\le k\le L} (\lambda_k,\, 0)^{\alpha_k} \prod_{L<j\le \ell} (\lambda_j,\, 0)^{\beta_j}.$$

By definition, if the natural number N is taken to be the common denominator of rationals γ_{jk}'s and γ'_{jk}'s, we find $(f^N,\, v'^N) \in (\mathcal{H}_\tau \times \mathbb{R}_{\ge 0}) \bigcap [\,\widehat{G}\,]$. Hence to show $(f^N,\, v^N) \in (\mathcal{H}_\tau \times \mathbb{R}_{\ge 0}) \bigcap [\,\widehat{G}\,]$, it is sufficient to prove $v = v'$.

Now assume $v = 0$. If $\nu_k(f) > 0$ for some $k \in J_Z(\xi) \cap \{1, 2, \ldots, L\}$, clearly we get $v' = 0$ because of $\delta_k = \nu_k(f) > 0$. Hence we assume $\nu_k(f) = 0$ for any $k \in J_Z(\xi) \cap \{1, 2, \ldots, L\}$. Then $v = 0$ implies either $\alpha_k > 0$ for some $1 \le k \le L$ or $\beta_j > 0$ for some $L < j \le \ell$ or $\alpha_k > 0$ for some $L < k \le m$ with $k \in J_Z(\xi)$. The last case implies $\delta_k > 0$ for the same k as τ_k only appears in g_k. Hence we have $v' = 0$ and we conclude that $(f^N,\, v^N) \in (\mathcal{H}_\tau \times \mathbb{R}_{\ge 0}) \bigcap [\,\widehat{G}\,]$ in this case.

Next assume $v \ne 0$. Then we have $\beta_j = 0$ for $L < j \le \ell$ and $\alpha_k = 0$ for any $1 \le k \le L$ or any $L < k$ with $k \in J_Z(\xi)$. Furthermore, we have $\nu_k(f) = 0$ for any $k \in J_Z(\xi) \cap \{1, 2, \ldots, L\}$ because $v \ne 0$ means $(f,\, v) \in [G]_=$. Hence, in particular, we obtain $\delta_k = \nu_k(f) = 0$ for $k \in J_Z(\xi)$ and

$$(f,\, v') = \prod_{1\le k\le m,\, k\notin J_Z(\xi)} (g_k,\, |\xi^{(k)}|)^{\delta_k},$$

which implies $v' \ne 0$. Then $v = v'$ follows from Lemma 2.15. This completes the proof. □

We can also show the following finiteness property.

Proposition 2.17. *$\mathcal{G}$ and $\widehat{\mathcal{G}}$ are finitely generated.*

Proof. We only show finiteness of $\mathcal{G}$ because the proof goes in the same way for the case of $\widehat{\mathcal{G}}$. We denote by $\pi : \mathcal{H}_\tau \times \mathbb{R}_{\ge 0} \to \mathcal{H}_\tau$ the projection which forgets v of $(f,\, v) \in \mathcal{H}_\tau \times \mathbb{R}_{\ge 0}$. Clearly we have $\pi(\mathcal{G}) = \pi([G]_\ge)$, and it follows from Gordan's lemma that $\pi([G]_\ge)$ is finitely generated.

Hence it suffices to show

$$\pi|_{\mathcal{G}} : \mathcal{G} \to \pi([G]_\ge)$$

is an isomorphism of semigroups. We make the inverse of $\pi|_{\mathcal{G}}$. Let $h \in \pi([G]_\ge)$. Then h has the unique form

$$h = (\varphi_1^{-1})^{\alpha_1} \cdots (\varphi_L^{-1})^{\alpha_L} \psi_{L+1}^{\alpha_{L+1}} \cdots \psi_m^{\alpha_m} \lambda_{L+1}^{\beta_{L+1}} \cdots \lambda_\ell^{\beta_\ell},$$

where $\beta_j \in \mathbb{Z}_{\ge 0}$ $(L < j \le \ell)$, $\alpha_k \in \mathbb{Z}_{\ge 0}$ for either $k \in J_Z(\zeta)$ or $1 \le k \le L$, and $\alpha_k \in \mathbb{Z}$ otherwise. As a matter of fact, the uniqueness comes from the condition (2.1) and the fact that τ_k $(k > L)$ is contained only in ψ_k.

Now we define

$$v := \begin{cases} 0 & \begin{array}{l}\text{if } \alpha_k > 0 \text{ for some } 1 \le k \le L \\ \text{or if } \beta_j > 0 \text{ for some } L < j \le \ell \\ \text{or if } \nu_k(h) > 0 \text{ for some } k \in J_Z(\xi) \cap \{1, 2, \dots, L\},\end{array} \\ \prod_{L<k\le m} n_k(\xi)^{\alpha_k} & \text{otherwise.} \end{cases}$$

Then we see $(h,\, v) \in \mathcal{G}$. It is easily checked that this correspondence from $\pi([G]_{\ge})$ to $\mathcal{G}$ gives a morphism of semigroup and becomes the inverse of $\pi|_{\mathcal{G}}$. □

The following lemma is useful through the paper.

Lemma 2.18. *We have the followings:*

1. *For $k \in J_Z(\xi)$ and $(f,\, v)$ in $\mathcal{G}$ or $\widehat{\mathcal{G}}$, we have $\nu_k(f) \ge 0$.*
2. *There exists $N \in \mathbb{N}$ such that $(\tau_k^N,\, 0)$ belongs to $\mathcal{G}$ and $\widehat{\mathcal{G}}$ $(1 \le k \le m)$.*

Proof. Since $\mathcal{G}$ and $\widehat{\mathcal{G}}$ are equivalent, it suffices to show the claims for $\widehat{\mathcal{G}}$.

The claim 1. comes immediately from the definition of $\widehat{G}$ because every pair in $\widehat{G}$ does not contain a negative power of the variable τ_k for $k \in J_Z(\xi)$.

Let us show the claim 2. If $\varphi_k = 1$, we have $\xi_k = 0$ by the assumption (2.3), and thus, $(\tau_k,\, 0) \in \widehat{\mathcal{G}}$. If $\varphi_k \ne 1$, by noticing $(\lambda_j,\, 0) \in \widehat{G}$ $(j = 1, \dots, \ell)$ and $(\tau_k/\varphi_k(\lambda),\, |\xi^{(k)}|) \in \widehat{G}$, we have $(\tau_k^N,\, 0) \in \widehat{\mathcal{G}}$ for sufficiently large $N \in \mathbb{N}$. This completes the proof. □

2.3 Families of cones

Let $p = (x^{(0)};\, \xi) = (x^{(0)}; \xi^{(1)}, \dots, \xi^{(m)})$ be a point in S_χ. Set

$$\begin{aligned} |x^{(*)}| &:= (|x^{(1)}|,\, |x^{(2)}|,\, \cdots,\, |x^{(L)}|,\, \cdots,\, |x^{(m)}|), \\ |x^{(*)}|' &:= (|x^{(1)}|,\, |x^{(2)}|,\, \cdots,\, |x^{(L)}|). \end{aligned} \tag{2.38}$$

Let H be a finite subset in $\mathcal{H}_\tau \times \mathbb{R}_{\ge 0}$. We denote by $C(\xi,\, H)$ a family of subsets of X in the form

$$\left\{ (x^{(0)}, x^{(1)}, \dots, x^{(m)}) \in X;\, \begin{array}{c} x^{(k)} \in W_k \quad (k = 1, \dots, m), \\ v - \epsilon < g\left(|x^{(*)}|\right) < v + \epsilon \\ \text{for } (g,\, v) \in H \end{array} \right\}, \tag{2.39}$$

where $\epsilon > 0$ and each W_k satisfies either 1. or 2. below.

1. W_k is an $\mathbb{R}^+$-conic open subset in $\mathbb{R}^{n_k} \setminus \{0\}$ containing the point $\xi^{(k)}$ if $k \notin J_Z(\xi)$.
2. W_k is just $\mathbb{R}^{n_k}$ if $k \in J_Z(\xi)$.

Lemma 2.19. *Assume that* $[H]$ *and* $\mathcal{G}$ *defined in* (2.29) *are equivalent. Then* $C(\xi, H)$ *is well-defined, and it is an open subset and bounded with respect to the variables* $x^{(1)}, \ldots, x^{(m)}$.

Proof. it follows from the claim 1. of Lemma 2.18 and the definition of W_k that, for any $(g, v) \in H$, the function $g\left(|x^{(*)}|\right)$ is well-defined and continuous on the region $(x^{(1)}, \ldots, x^{(m)}) \in W_1 \times \cdots \times W_m$. Hence $C(\xi, H)$ is an open subset. Boundness is a consequence of the claim 2. of the same lemma. □

Then we set

$$C(\xi) = \bigcup_{H} C(\xi, H), \tag{2.40}$$

where H runs through the family of finite subsets in $\mathcal{H}_\tau \times \mathbb{R}_{\geq 0}$ such that $[H]$ and $\mathcal{G}$ are equivalent. Let $\pi : S_\chi \to M = M_1 \cap \cdots \cap M_\ell$ be the canonical projection.

Definition 2.20. *We say that an open subset* $V \subset X$ *is a germ of a multi-cone at* $p \in S_\chi$ *if there exist an open neighborhood* Ω *of* $\pi(p)$ *in* X *and* $W \in C(\xi)$ *satisfying*

$$(W \cap \Omega) \subset V. \tag{2.41}$$

Remark 2.21. *As the semigroup* $\mathcal{G}$ *is finitely generated, we can take generators of* $\mathcal{G}$ *as* H. *Then it is easy to see that* $C(\xi, H)$ *becomes a cofinal family of* $C(\xi)$. *Furthermore, since* $\mathcal{G}$ *and* $\widehat{\mathcal{G}}$ *are equivalent, we can obtain the same family of multi-cones if we replace* $\mathcal{G}$ *with* $\widehat{\mathcal{G}}$ *in the above definition.*

Remember $q := \#(J_Z(\xi) \cap \{1, \ldots, L\})$. We now give the explicit method to obtain a finite set F^q in $\mathcal{H}_\tau \times \mathbb{R}_{\geq 0}$ for which $C(\xi, F^q)$ gives a cofinal family of $C(\xi)$.

Let F be a finite set in $\mathcal{H}_{\tau,\lambda} \times \mathbb{R}_{\geq 0}$ and $k \in \{1, 2, \ldots, L\}$. We will consider an operation $\mathcal{L}_k$ on F which generates a finite subset in $\mathcal{H}_{\tau,\lambda} \times \mathbb{R}_{\geq 0}$: The subset $\mathcal{L}_k(F) \subset \mathcal{H}_{\tau,\lambda} \times \mathbb{R}_{\geq 0}$ consists of the following elements:

F1. (f, v) for $(f, v) \in F$ with $\nu_k(f) = 0$.

F2. $(f, 0)$ for $(f, v) \in F$ with $\nu_k(f) > 0$.

F3. $(f^a g^b, v^a w^b)$ where (f, v) and (g, w) in F with $\nu_k(f) > 0$ and $\nu_k(g) < 0$, and the natural numbers a and b are taken to be prime to each other with $a|\nu_k(f)| = b|\nu_k(g)|$.

The following lemma is an easy consequence of the above operations.

Lemma 2.22. *For any* $(f, v) \in \mathcal{L}_k(F)$, *we have* $\nu_k(f) \geq 0$. *Furthermore, for* $(f, v) \in \mathcal{L}_k(F)$ *with* $v \neq 0$, *we have* $\nu_k(f) = 0$, *i.e.,* f *does not depend on* τ_k.

Let $j \in \{1, 2, \ldots, \ell\}$. We also introduce the similar operation $\mathcal{L}_j^\lambda(F)$ with respect to the variable λ_j. The subset $\mathcal{L}_j^\lambda(F) \subset \mathcal{H}_{\tau,\lambda} \times \mathbb{R}_{\geq 0}$ consists of the following elements:

F1. $(f,\, v)$ for $(f,\, v) \in F$ with $\nu_j^\lambda(f) = 0$.

F3. $(f^a g^b,\, v^a w^b)$ where (f, v) and (g, w) in F with $\nu_j^\lambda(f) > 0$ and $\nu_j^\lambda(g) < 0$, and the natural numbers a and b are taken to be prime to each other with $a|\nu_j^\lambda(f)| = b|\nu_j^\lambda(g)|$.

Note that, in this case, there is no counterpart of the operation F2 in $\mathcal{L}_k$. Hence, unlike the operation $\mathcal{L}_k$, the variable λ_j is eliminated completely by the operation $\mathcal{L}_j^\lambda$.

Recall the finite subset $G \subset \mathcal{H}_{\tau,\lambda} \times \mathbb{R}_{\geq 0}$ defined by (2.24) or (2.25) when ξ is not normalized. Then, we apply the operation $\mathcal{L}_j^\lambda$'s $(L < j \leq \ell)$ successively to G, we obtain the finite subset F^0 whose elements are independent of the variables λ because an element of G is independent of the variables λ' and the operation $\mathcal{L}_j^\lambda$ eliminates the variable λ_j $(L < j \leq \ell)$, that is,

$$F^0 := (\mathcal{L}_\ell^\lambda \circ \cdots \circ \mathcal{L}_{L+1}^\lambda)(G) \ \subset \mathcal{H}_\tau \times \mathbb{R}_{\geq 0}. \tag{2.42}$$

Next we apply the operations $\mathcal{L}_{k_1}, \ldots, \mathcal{L}_{k_q}$ successively to F^0, we get the subset $F^q \subset \mathcal{H}_\tau \times \mathbb{R}_{\geq 0}$, i.e.,

$$F^q := (\mathcal{L}_{k_q} \circ \cdots \circ \mathcal{L}_{k_1})(F^0) \ \subset \mathcal{H}_\tau \times \mathbb{R}_{\geq 0}, \tag{2.43}$$

where $q = \#(J_Z(\xi) \cap \{1, 2, \ldots, L\})$ and $J_Z(\xi) \cap \{1, 2, \ldots, L\} = \{k_1, \ldots, k_q\}$ with $1 \leq k_1 < \cdots < k_q \leq L$.

Remark 2.23. *For convenience, we also set*

$$F^{0,r} := (\mathcal{L}_r^\lambda \circ \cdots \circ \mathcal{L}_{L+1}^\lambda)(G) \ \subset \mathcal{H}_{\tau,\lambda} \times \mathbb{R}_{\geq 0} \qquad (L < r \leq \ell), \tag{2.44}$$

and

$$F^s := (\mathcal{L}_{k_s} \circ \cdots \circ \mathcal{L}_{k_1})(F^0) \subset \mathcal{H}_\tau \times \mathbb{R}_{\geq 0} \qquad (1 \leq s \leq q). \tag{2.45}$$

Remark 2.24. *If μ_X is non-degenerate, as $L = \ell$ holds, we get $F^0 = G$, i.e., an element of G is independent of the variables λ. Hence we have*

$$F^q := (\mathcal{L}_{k_q} \circ \cdots \circ \mathcal{L}_{k_1})(G). \tag{2.46}$$

On the other hand, when $p = (x^{(0)};\, \xi) \in S_X$ is outside fixed points, by Corollary 2.5, we can take coordinates blocks so that the conditions 1. and 2. of the corollary hold. Then, under such coordinates blocks, since $J_Z(\xi) \cap \{1, 2, \ldots, L\}$ becomes empty, we have $q = 0$ which implies

$$F^q = F^0 = (\mathcal{L}_\ell^\lambda \circ \cdots \circ \mathcal{L}_{L+1}^\lambda)(G).$$

In particular, if μ_X is non-degenerate and p is outside fixed points, under the above coordinates blocks, we have

$$F^q = F^0 = G.$$

The following lemma easily follows from the previous lemma for $k \in J_Z(\xi)$ with $k \leq L$ and from the explicit form of (ψ_k, n_k) for $k \in J_Z(\xi)$ with $k > L$.

Lemma 2.25. *For any $(f,v) \in F^q$ and $k \in J_Z(\xi)$, we have $\nu_k(f) \geq 0$. Furthermore, for $(f,v) \in F^q$ with $v \neq 0$ and $k \in J_Z(\xi)$, we have $\nu_k(f) = 0$.*

Example 2.26. *In $\mathbb{R}^3$ let us consider the family $\chi = \{M_1, M_2\}$, with $M_1 = \{x_1 = x_2 = 0\}$, $M_2 = \{x_2 = x_3 = 0\}$. The matrix A is*

$$\begin{pmatrix} 1 & 1 & 0 \\ 0 & 1 & 1 \end{pmatrix}$$

and $\varphi_1^{-1}(\tau) = \tau_1$, $\varphi_2^{-1}(\tau) = \dfrac{\tau_2}{\tau_1}$, $\psi_3(\tau) = \dfrac{\tau_1\tau_3}{\tau_2}$. We have

$$F^0 = Q\left(\left\{(\tau_1, 0), \left(\frac{\tau_2}{\tau_1}, 0\right), \left(\frac{\tau_1\tau_3}{\tau_2}, |\xi_3|\right)\right\}\right).$$

In the following, except $(1,1)$ (we omit it), all the couples (f,v) are with $v = 0$, so we omit v for short.

- *If $\xi_3 = 0$, $\xi_1, \xi_2 \neq 0$, $F^0 = \left\{\tau_1, \dfrac{\tau_2}{\tau_1}, \dfrac{\tau_1\tau_3}{\tau_2}\right\}$.*
- *If $\xi_1 = 0$, $\xi_2, \xi_3 \neq 0$, $F^1 = \left\{\tau_1, \tau_2, \tau_3, \dfrac{\tau_1\tau_3}{\tau_2}, \dfrac{\tau_2}{\tau_3}\right\}$.*
- *If $\xi_2 = 0$, $\xi_1, \xi_3 \neq 0$, $F^1 = \left\{\tau_1, \dfrac{\tau_2}{\tau_1}, \tau_3, \dfrac{\tau_2}{\tau_1\tau_3}\right\}$.*
- *If $\xi_1 \neq 0$, $\xi_2, \xi_3 = 0$, $F^2 = \left\{\tau_1, \tau_2, \tau_3, \dfrac{\tau_2}{\tau_1}\right\}$.*
- *If $\xi_2 \neq 0$, $\xi_1, \xi_3 = 0$, $F^2 = \left\{\tau_1, \tau_2, \tau_3, \dfrac{\tau_1\tau_3}{\tau_2}\right\}$.*
- *If $\xi_3 \neq 0$, $\xi_1, \xi_2 = 0$, $F^2 = \left\{\tau_1, \tau_2, \tau_3, \dfrac{\tau_2}{\tau_3}\right\}$.*

Example 2.27. *In $\mathbb{R}^4$, let us consider the family $\chi = \{M_1, M_2\}$, with $M_1 = \{x_1 = x_2 = x_4 = 0\}$, $M_2 = \{x_2 = x_3 = 0\}$. The matrix A is*

$$\begin{pmatrix} 1 & 1 & 0 & 1 \\ 0 & 1 & 1 & 0 \end{pmatrix}.$$

Then $\varphi_1^{-1}(\tau) = \tau_1$, $\varphi_2^{-1}(\tau) = \dfrac{\tau_2}{\tau_1}$, $\psi_3(\tau) = \dfrac{\tau_1\tau_3}{\tau_2}$, $\psi_4(\tau) = \dfrac{\tau_4}{\tau_1}$. We have

$$F^0 = Q\left(\left\{(\tau_1, 0), \left(\frac{\tau_2}{\tau_1}, 0\right), \left(\frac{\tau_1\tau_3}{\tau_2}, |\xi_3|\right), \left(\frac{\tau_4}{\tau_1}, |\xi_4|\right)\right\}\right).$$

Suppose that $\xi_1, \xi_2 \neq 0$. Let us consider the families F^1 and F^2.

- *If $\xi_3 = 0$, $\xi_1, \xi_2, \xi_4 \neq 0$, we have*

$$F^1 = \left\{(\tau_1, 0), \left(\frac{\tau_2}{\tau_1}, 0\right), \left(\frac{\tau_1\tau_3}{\tau_2}, 0\right), \left(\frac{\tau_4}{\tau_1}, |\xi_4|\right), \left(\frac{\tau_1}{\tau_4}, \frac{1}{|\xi_4|}\right), (1,1)\right\}.$$

- *If* $\xi_4 = 0$, $\xi_1, \xi_2, \xi_3 \neq 0$, *we have*

$$F^1 = \left\{(\tau_1, 0), \left(\frac{\tau_2}{\tau_1}, 0\right), \left(\frac{\tau_1\tau_3}{\tau_2}, |\xi_3|\right), \left(\frac{\tau_2}{\tau_1\tau_3}, \frac{1}{|\xi_3|}\right), \left(\frac{\tau_4}{\tau_1}, 0\right), (1,1)\right\}.$$

- *If* $\xi_3, \xi_4 = 0$, $\xi_1, \xi_2 \neq 0$, *we have*

$$F^2 = \left\{(\tau_1, 0), \left(\frac{\tau_2}{\tau_1}, 0\right), \left(\frac{\tau_1\tau_3}{\tau_2}, 0\right), \left(\frac{\tau_4}{\tau_1}, 0\right), (1,1)\right\}.$$

F^q is not a family of generators of $\mathcal{G}$ in general, however, it enjoys the following good property from a geometrical point of view.

Proposition 2.28. *1. We have* $[F^q] \subset \mathcal{G}$. *We also have* $\mathrm{Q}(F^s) = F^s$ *for any* $1 \le s \le q$ *and* $\mathrm{Q}(F^{0,r}) = F^{0,r}$ *for* $L < r \le \ell$.

2. $[F^q]$ *and* $\mathcal{G}$ *are equivalent. In particular, the radical of* $[F^q]$ *is* $\mathcal{G}$. *That is, there exists* $N \in \mathbb{N}$ *such that* $(f,\, v)^N \in [F^q]$ *for any* $(f,\, v) \in \mathcal{G}$.

Proof. These claims can be proved by the induction on the number of operations $\mathcal{L}_j^\lambda$'s and $\mathcal{L}_{k_i}$'s. Since argument used in the inductive step on the number of the operations $\mathcal{L}_j^\lambda$'s is the same as that for the operations $\mathcal{L}_{k_i}$'s, for simplicity, we show the claims only for the case $L = \ell$, i.e., $F^0 = G$.

It is easy to see that, for a subset $A \subset \mathcal{H}_{\tau,\lambda} \times \mathbb{R}_{\ge 0}$, we have

$$(\mathrm{Q} \circ \mathcal{L}_k \circ \mathrm{Q})(A) = (\mathcal{L}_k \circ \mathrm{Q})(A).$$

In what follows, we denote by $A_{>\tilde{q}}$, $\mathcal{G}^{\tilde{q}}$ and etc. the corresponding objects when $q = \tilde{q}$.

Let us show the claim 1. by the induction on q. When $q = 0$, the claim clearly holds. Suppose that the claim is true for $q = \tilde{q}$, and we will show the claim when $q = \tilde{q} + 1$. We have

$$\begin{aligned} F^{\tilde{q}+1} &= \mathcal{L}_{k_{\tilde{q}+1}}(F^{\tilde{q}}) = (\mathcal{L}_{k_{\tilde{q}+1}} \circ \mathrm{Q})(F^{\tilde{q}}) = (\mathrm{Q} \circ \mathcal{L}_{k_{\tilde{q}+1}} \circ \mathrm{Q})(F^{\tilde{q}}) \\ &= (\mathrm{Q} \circ \mathcal{L}_{k_{\tilde{q}+1}})(F^{\tilde{q}}) = \mathrm{Q}(F^{\tilde{q}+1}), \end{aligned}$$

which show the first claim of 1. for $q = \tilde{q} + 1$. Since we have

$$F^{\tilde{q}+1} = \mathcal{L}_{k_{\tilde{q}+1}}(F^{\tilde{q}}) \subset \mathcal{L}_{k_{\tilde{q}+1}}(\mathcal{G}^{\tilde{q}})$$

by the induction hypothesis and since we can easily confirm $\mathcal{L}_{k_{\tilde{q}+1}}(\mathcal{G}^{\tilde{q}}) \subset \mathcal{G}^{\tilde{q}+1}$, we have obtained $[F^{\tilde{q}+1}] \subset \mathcal{G}^{\tilde{q}+1}$, which shows the second claim of 1. to be true for $q = \tilde{q} + 1$. Hence, by the induction, we have shown the claim 1.

Next we will show the claim 2. by the induction on $q \ge 0$. Assume first $q = 0$. In this case, $\mathcal{G} = [G] = [F^0]$ holds, and hence, the claim 2. is true.

Suppose that the claim 2. is true for $q = \tilde{q} - 1 \geq 0$. We will show the claim 2. for $q = \tilde{q}$. Let $(f, v) \in \mathcal{G}^{\tilde{q}}$ and assume $F^{\tilde{q}-1} = \{(h_r, v_r)\}_r$.

Let us first consider the case $v \neq 0$. In this case, we have $(f, v) \in [G]_{=\tilde{q}}$, and hence, $(f, v) \in \mathcal{G}^{\tilde{q}-1}$ also holds. It follows from induction hypothesis that there exist $N_{\tilde{q}-1} \in \mathbb{N}$ and a family $\{\alpha_r\}_r$ of non-negative integers such that

$$(f, v)^{N_{\tilde{q}-1}} = \prod_r (h_r, v_r)^{\alpha_r}.$$

Then $(f, v) \in [G]_{=\tilde{q}}$ implies

$$\sum_r \alpha_r \nu_{k_{\tilde{q}}}(h_r) = 0.$$

Now let us consider the following procedure:

Choose an index r^* such that $|\alpha_{r^*}\nu_{k_{\tilde{q}}}(h_{r^*})|$ is minimum in the set of non-zero $|\alpha_r \nu_{k_{\tilde{q}}}(h_r)|$'s, and then, find an index r' and a positive rational number $\beta_{r'}$ such that

$$0 < \beta_{r'} \leq \alpha_{r'}, \quad \alpha_{r^*}\nu_{k_{\tilde{q}}}(h_{r^*}) + \beta_{r'}\nu_{k_{\tilde{q}}}(h_{r'}) = 0.$$

Note that such an index r' and $\beta_{r'}$ necessarily exist by the choice of r^*. By repeated applications of the procedure described above, we can find a finite number of $(g_r, w_r) \in F^{\tilde{q}}$ with $\nu_{k_{\tilde{q}}}(g_r) = 0$ and positive rational numbers β_r such that

$$(f, v)^{N_{\tilde{q}-1}} = \prod_r (g_r, w_r)^{\beta_r},$$

where each g_r is generated by the operations F1 or F3.

Next we consider the case $v = 0$. By induction hypothesis, there exist $N_{\tilde{q}-1} \in \mathbb{N}$ and a family $\{\alpha_r\}_r$ of non-negative integers such that we have either

$$(f, v)^{N_{\tilde{q}-1}} = \prod_r (h_r, v_r)^{\alpha_r} * (1, 0), \qquad \sum_r \alpha_r \nu_{k_{\tilde{q}}}(h_r) > 0$$

if $\nu_{k_i}(f) = 0$ for $1 \leq i \leq \tilde{q} - 1$ and $\nu_{k_{\tilde{q}}}(f) > 0$ hold, or

$$(f, v)^{N_{\tilde{q}-1}} = \prod_r (h_r, v_r)^{\alpha_r}, \qquad \sum_r \alpha_r \nu_{k_{\tilde{q}}}(h_r) \geq 0$$

otherwise. In both the above cases, by applying the same argument, we can find a finite number of $(g_r, w_r) \in F^{\tilde{q}}$ with $\nu_{k_{\tilde{q}}}(g_r) = 0$, $(\tilde{g}_s, 0) \in F^{\tilde{q}}$ with $\nu_{k_{\tilde{q}}}(\tilde{g}_s) > 0$, positive rational numbers β_r and $\tilde{\beta}_s$ such that

$$(f, v)^{N_{\tilde{q}-1}} = \prod_r (g_r, w_r)^{\beta_r} \prod_s (\tilde{g}_s, 0)^{\tilde{\beta}_s},$$

where each g_r is generated by the operation F1 or F3 and each $\tilde{g}_s$ is a consequence of the operation F2.

As a conclusion, for each $(f, v) \in \mathcal{G}^{\tilde{q}}$, there exists $N \in \mathbb{N}$ such that $(f, v)^N$ belongs to $[F^{\tilde{q}}]$. Furthermore, since $\mathcal{G}^{\tilde{q}}$ is finitely generated, we can take such an N uniformly. Therefore the claim 2. is true for $q = \tilde{q}$, and thus, it is true for all q. This completes the proof. □

As an immediate corollary, we obtain:

Corollary 2.29. *$C(\xi, F^q)$ gives a cofinal family of $C(\xi)$.*

Example 2.30. *Now let us consider the following simple example. Let $m = 3$ and $\ell = 2$ where the action μ is defined by*

$$(x^{(0)}, \lambda_1 x^{(1)}, \lambda_2 x^{(2)}, \lambda_1\lambda_2 x^{(3)}) \qquad (\lambda = (\lambda_1, \lambda_2) \in (\mathbb{R}^+)^2).$$

Hence we have

$$\varphi_1(\lambda) = \lambda_1,\ \varphi_2(\lambda) = \lambda_2,\ \varphi_3(\lambda) = \lambda_1\lambda_2$$

and

$$\varphi_1^{-1}(\tau) = \tau_1,\ \varphi_2^{-1}(\tau) = \tau_2.$$

Take $p = (0;\, 0, 0, \xi^{(3)}) \in S_\chi$ with $\xi^{(3)} \neq 0$. Then the initial set F^0 is given by

$$\left\{(\tau_1, 0),\ (\tau_2, 0),\ \left(\frac{\tau_3}{\tau_1\tau_2}, |\xi^{(3)}|\right),\ \left(\frac{\tau_1\tau_2}{\tau_3}, \frac{1}{|\xi^{(3)}|}\right)\right\}.$$

Therefore F^1 is given by

$$\left\{(\tau_1, 0),\ (\tau_2, 0),\ \left(\frac{\tau_1\tau_2}{\tau_3}, 0\right),\ \left(\frac{\tau_3}{\tau_2}, 0\right),\ (1, 1)\right\}.$$

Then F^2 is

$$\left\{(\tau_1, 0),\ (\tau_2, 0),\ \left(\frac{\tau_1\tau_2}{\tau_3}, 0\right),\ (\tau_3, 0),\ (1, 1)\right\}.$$

Hence the most important inequality of a multi-cone in this case is

$$0 - \epsilon < \frac{|x^{(1)}||x^{(2)}|}{|x^{(3)}|} < 0 + \epsilon \iff |x^{(1)}||x^{(2)}| < \epsilon |x^{(3)}|.$$

As a conclusion, a cofinal family of multi-cones is given by

$$\left\{\begin{array}{c} x^{(3)} \in W_3, \\ (x^{(0)}, x^{(1)}, x^{(2)}, x^{(3)});\ |x^{(1)}| < \epsilon,\ |x^{(2)}| < \epsilon, |x^{(3)}| < \epsilon, \\ |x^{(1)}||x^{(2)}| < \epsilon |x^{(3)}| \end{array}\right\},$$

where $\epsilon > 0$ and W_3 is a proper convex cone containing the direction $\xi^{(3)}$.

We have the following characterization of the multi-normal cone by using multi-cones thus defined.

Theorem 2.31. *Let $p = (x^{(0)}; \xi) \in S_{\mathcal{X}}$ and Z a subset in X. Then the following conditions are equivalent.*

i) $p \notin C_{\mathcal{X}}(Z)$.

ii) There exist a $V \in C(\xi)$ and an open neighborhood U of $\pi(p)$ in X such that $Z \cap V \cap U = \emptyset$.

iii) There exist a $V \in C(\xi, F^q)$ and an open neighborhood U of $\pi(p)$ in X such that $Z \cap V \cap U = \emptyset$.

To prove the theorem, it suffices to show the following equivalence:

1. $p \in C_{\mathcal{X}}(Z)$.
2. $Z \cap V \cap U$ is non-empty for any $V \in C(\xi)$ and any open neighborhood U of $\pi(p)$.
3. $Z \cap V \cap U$ is non-empty for any $V \in C(\xi, F^q)$ and any open neighborhood U of $\pi(p)$.

We will prove these equivalences step by step in several lemmas below. Clearly the implication from 2. to 3 holds. Hence we first show the implication from 1. to 2.

Lemma 2.32. *If $p \in C_{\mathcal{X}}(Z)$, then $Z \cap V \cap U$ is non-empty for any $V \in C(\xi)$ and any open neighborhood U of $\pi(p)$.*

Proof. Let

$$p_i = (\tilde{x}_i, t_i) = (\tilde{x}_i^{(0)}, \tilde{x}_i^{(1)}, \ldots, \tilde{x}_i^{(m)}, t_{1,i}, \ldots, t_{\ell,i}) \in \widetilde{X} \qquad (i = 1, 2, \ldots)$$

be a sequence satisfying that $\widetilde{p}(p_i) \in Z$ and

$$\tilde{x}_i^{(0)} \to x^{(0)}, \quad \tilde{x}_i^{(k)} \to \xi^{(k)}\ (k = 1, \ldots, m), \quad t_{j,i} \to 0\ (j = 1, \ldots, \ell),$$

when $i \to \infty$. Set

$$x_i = (x_i^{(0)}, x_i^{(1)}, \ldots, x_i^{(m)}) := \tilde{p}(p_i) \in Z \subset X.$$

By Proposition 2.13, it is enough to prove, for any $(f, v) \in (\mathcal{H}_\tau \times \mathbb{R}_{\geq 0}) \cap [\,\widehat{G}\,]$,

$$f(|x_i^{(1)}|, \ldots, |x_i^{(m)}|) \to v \qquad (i \to \infty). \tag{2.47}$$

Set $g_k := \tau_k / \varphi_k(\lambda)$ $(k = 1, 2, \ldots, m)$. Then, by definition of $\widehat{G}$, there exist $\alpha_k, \beta_j \in \mathbb{Z}$ such that

$$f(\tau) = \prod_{1 \leq k \leq m} g_k(\tau, \lambda)^{\alpha_k} \prod_{1 \leq j \leq \ell} \lambda_j^{\beta_j},$$

where $\beta_j \geq 0$ and $\alpha_k \geq 0$ if $k \in J_Z(\xi)$. Now, by putting $\tau_k = |x_i^{(k)}|$ and $\lambda_j = t_{j,i}^{\sigma_A}$ into f, we have

$$f(|x_i^{(*)}|) = \prod_{1 \leq k \leq m} g_k(|x_i^{(*)}|, t_i^{\sigma_A})^{\alpha_k} \prod_{1 \leq j \leq \ell} t_{j,i}^{\sigma_A \beta_j},$$

where $t_i^{\sigma_A} := (t_{1,i}^{\sigma_A}, \ldots, t_{\ell,i}^{\sigma_A})$. Note that

$$g_k(|x_i^{(*)}|, t_i^{\sigma_A}) = \frac{\varphi_k(t_i)^{\sigma_A}|\tilde{x}_i^{(k)}|}{\varphi_k(t_i^{\sigma_A})} = |\tilde{x}_i^{(k)}| \to |\xi^{(k)}| \qquad (k = 1, 2, \ldots, m)$$

holds when $i \to \infty$.

If $v = 0$, then we have either $\beta_j > 0$ for some j or $\alpha_k > 0$ for some $k \in J_Z(\xi)$. Hence we get

$$f(|x_i^{(*)}|) = \prod_{1 \le k \le m} g_k(|x_i^{(*)}|, t_i^{\sigma_A})^{\alpha_k} \prod_{1 \le j \le \ell} t_{j,i}^{\sigma_A \beta_j} \to 0 = v.$$

If $v \neq 0$, then we have $\beta_j = 0$ and $\alpha_k = 0$ for any $k \in J_Z(\xi)$. Therefore we get

$$f(|x_i^{(*)}|) = \prod_{1 \le k \le m, k \notin J_Z(\xi)} g_k(|x_i^{(*)}|, t_i^{\sigma_A})^{\alpha_k} \to \prod_{1 \le k \le m, k \notin J_Z(\xi)} |\xi^{(k)}|^{\alpha_k}.$$

It follows from Lemma 2.15 that the last term is equal to v. This completes the proof. □

Let us show the implication from 3. to 1. We need to prepare several lemmas.

Lemma 2.33. *Let $a_1, \ldots, a_m \in \mathbb{R}^\times$ be non-zero real numbers and $v_1, \ldots, v_m \in \mathbb{R}_{\ge 0}$ non-negative real numbers, and let $\{\kappa_{k,i}\}_{i=1}^\infty$ $(k = 1, \ldots, m)$ be m-sequences of positive real numbers. Then the following two conditions are equivalent.*

1. *There exists a sequence $\{\epsilon_i\}_{i=1}^\infty$ of positive real numbers satisfying, when $i \to \infty$,*

$$\epsilon_i \to 0, \qquad (\epsilon_i)^{-a_k}\kappa_{k,i} \to v_k \quad (k = 1, \ldots, m).$$

2. *The m-sequences satisfy, when $i \to \infty$,*

$$\begin{cases} \kappa_{p,i} \to 0 & (p \in O_+ \cup N_+), \\ 1/\kappa_{p,i} \to 0 & (p \in N_-), \\ (\kappa_{p,i})^{-a_q}(\kappa_{q,i})^{a_p} \to (v_{p,i})^{-a_q}(v_{q,i})^{a_p} & \begin{pmatrix} p \in O_+ \cup N_+ \ \textit{and} \\ q \in O_- \cup N_- \end{pmatrix}, \\ (\kappa_{p,i})^{|a_q|}(\kappa_{q,i})^{-|a_p|} \to (v_{p,i})^{|a_q|}(v_{q,i})^{-|a_p|} & \begin{pmatrix} p \in O_+ \cup N_+, q \in N_+ \ \textit{or} \\ p \in O_- \cup N_-, q \in N_- \end{pmatrix}, \end{cases}$$

where we set $O_+ = \{k;\, v_k = 0,\, a_k > 0\}$, $O_- = \{k;\, v_k = 0,\, a_k < 0\}$, $N_+ = \{k;\, v_k > 0,\, a_k > 0\}$ and $N_- = \{k;\, v_k > 0,\, a_k < 0\}$.

Proof. The implication from 1. to 2. is trivial. Let us show the converse implication.

First assume that $N_+ \cup N_-$ is empty. By considering $(\kappa_{j,m})^{1/|a_j|}$, we may assume $a_j = \pm 1$. Determine two sequences of positive real numbers

$$c_m := \max_{p \in O_+} \kappa_{p,m}, \quad d_m := \max_{q \in O_-} \kappa_{q,m}.$$

Then, by the conditions, we have

$$1/c_m \to \infty, \quad d_m \prec 1/c_m.$$

Here $\alpha_m \prec \beta_m$ means $\alpha_m/\beta_m \to 0$ $(m \to \infty)$. Now we can find a sequence $\{\epsilon_m\}$ so that

$$\epsilon_m \to 0, \quad d_m \prec 1/\epsilon_m \prec 1/c_m,$$

which satisfies the required condition.

Now assume $N_+ \cup N_-$ to be non-empty. If N_+ is non-empty, we take some $p \in N_+$ and set $\epsilon_i := (\kappa_{p,i}/v_p)^{1/a_p}$. It is easy to see that 1. holds for $\{\epsilon_i\}$. The other case can be shown in the same way. □

Then we introduce the modified operations $\tilde{\mathcal{L}}_k$ of $\mathcal{L}_k$ and $\tilde{\mathcal{L}}_j^\lambda$ of $\mathcal{L}_j^\lambda$, which generate conditions appearing in the above lemma. For a finite subset F in $\mathcal{H}_{\tau,\lambda} \times \mathbb{R}_{\geq 0}$, we define some subsets $F_{0,>}$, $F_{0,<}$, $F_{\times,>}$ and $F_{\times,<}$ of F by

$$\begin{aligned} F_{0,>} &:= \{(f,v) \in F;\, v = 0,\, \nu_k(f) > 0\}, \\ F_{0,<} &:= \{(f,v) \in F;\, v = 0,\, \nu_k(f) < 0\}, \\ F_{\times,>} &:= \{(f,v) \in F;\, v > 0,\, \nu_k(f) > 0\}, \\ F_{\times,<} &:= \{(f,v) \in F;\, v > 0,\, \nu_k(f) < 0\}. \end{aligned}$$

Then, the set $\tilde{\mathcal{L}}_k(F)$ consists of the following elements:

$\tilde{\mathrm{F}}1$. $(f,\, v)$ for $(f,\, v) \in F$ with $\nu_k(f) = 0$.

$\tilde{\mathrm{F}}2$. $(f,\, 0)$ where $(f,v) \in F_{0,>} \cup F_{\times,>}$ or $f = 1/g$ for some $(g,v) \in F_{\times,<}$.

$\tilde{\mathrm{F}}3$. $(f^a g^b,\, v^a w^b)$ where we take a combination of pairs (f,v) and (g,w) in sets with indices $>$ and $<$ respectively, that is, $(f,v) \in F_{0,>} \cup F_{\times,>}$ and $(g,w) \in F_{0,<} \cup F_{\times,<}$. And see below for $a, b \in \mathbb{N}$.

$\tilde{\mathrm{F}}4$. $(f^a g^{-b},\, v^a w^{-b})$ where a pair satisfies

(a) $(f,v) \in F_{0,>} \cup F_{\times,>}$ and $(g,w) \in F_{\times,>}$.

(b) $(f,v) \in F_{0,<} \cup F_{\times,<}$ and $(g,w) \in F_{\times,<}$.

Here the natural numbers a and b appearing in $\tilde{\mathrm{F}}3$ and $\tilde{\mathrm{F}}4$ are taken to be prime to each other with $a|\nu_k(f)| = b|\nu_k(g)|$.

The modified operation $\tilde{\mathcal{L}}_j^\lambda$ is defined in the similar way: We first define subsets $F_{0,>}$, $F_{0,<}$, $F_{\times,>}$ and $F_{\times,<}$ of F by replacing $\nu_k(f)$ with $\nu_j^\lambda(f)$ in the above definitions of corresponding subsets for $\tilde{\mathcal{L}}_k$. Then, the set $\tilde{\mathcal{L}}_j^\lambda(F)$ consists of the following elements:

F̃1. $(f,\, v)$ for $(f,\, v) \in F$ with $\nu_j^\lambda(f) = 0$.

F̃2. $\left(\lambda_j^a f^b,\, 0\right)$ where $(f, v) \in F_{0,<} \cup F_{\times,<}$ or $f = 1/g$ for some $(g, v) \in F_{\times,>}$, and the natural numbers a and b are taken to be prime to each other with $|\nu_j^\lambda(f)| = a/b$.

F̃3. $(f^a g^b,\, v^a w^b)$ where we take a combination of pairs (f, v) and (g, w) in sets with indices $>$ and $<$ respectively, that is, $(f, v) \in F_{0,>} \cup F_{\times,>}$ and $(g, w) \in F_{0,<} \cup F_{\times,<}$, and see below for $a, b \in \mathbb{N}$.

F̃4. $(f^a g^{-b},\, v^a w^{-b})$ where a pair satisfies

(a) $(f, v) \in F_{0,>} \cup F_{\times,>}$ and $(g, w) \in F_{\times,>}$.

(b) $(f, v) \in F_{0,<} \cup F_{\times,<}$ and $(g, w) \in F_{\times,<}$.

Here the natural numbers a and b appearing in F̃3 and F̃4 are taken to be prime to each other with $a|\nu_j^\lambda(f)| = b|\nu_j^\lambda(g)|$.

Remark 2.34. *The above F̃1, F̃3 and F̃4 are the same as those in $\tilde{\mathcal{L}}_k$, where we just replace $\nu_k(f)$ and $\nu_k(g)$ with $\nu_j^\lambda(f)$ and $\nu_j^\lambda(g)$ in their definitions, respectively. While F̃2 in $\tilde{\mathcal{L}}_j^\lambda$ is different from the corresponding one in $\tilde{\mathcal{L}}_k$.*

Then we can easily confirm the following lemma:

Lemma 2.35. *1. Assume $F = \mathrm{Q}(F)$. Then we have $\mathcal{L}_k(F) = \tilde{\mathcal{L}}_k(F)$.*

2. Assume $F = \mathrm{Q}(F)$ and $(\lambda_j,\, 0) \in F$. Then we have $\mathcal{L}_j^\lambda(F) = \tilde{\mathcal{L}}_j^\lambda(F)$.

It follows from the lemma and facts

$$G = \mathrm{Q}(G), \quad Q \circ \mathcal{L}_j^\lambda \circ Q = \mathcal{L}_j^\lambda \circ Q, \quad Q \circ \mathcal{L}_{k_i} \circ Q = \mathcal{L}_{k_i} \circ Q$$

and

$$Q \circ \tilde{\mathcal{L}}_j^\lambda \circ Q = \tilde{\mathcal{L}}_j^\lambda \circ Q, \quad Q \circ \tilde{\mathcal{L}}_{k_i} \circ Q = \tilde{\mathcal{L}}_{k_i} \circ Q$$

that we get

$$F^0 = (\mathcal{L}_\ell^\lambda \circ \cdots \circ \mathcal{L}_{L+1}^\lambda)(G) = (\tilde{\mathcal{L}}_\ell^\lambda \circ \cdots \circ \tilde{\mathcal{L}}_{L+1}^\lambda)(G)$$

and

$$F^q = (\mathcal{L}_{k_q} \circ \cdots \circ \mathcal{L}_{k_1})(F^0) = (\tilde{\mathcal{L}}_{k_q} \circ \cdots \circ \tilde{\mathcal{L}}_{k_1})(F^0).$$

As a conclusion, we may consider F^q to be generated by the modified operations $\tilde{\mathcal{L}}_j^\lambda$'s $(L < j \leq m)$ and $\tilde{\mathcal{L}}_{k_i}$'s $(1 \leq i \leq q)$.

Proof. Now we are ready to show the implication from 3. to 1. We may assume $x^{(0)} = 0$. We can find a sequence in Z

$$x_i = (x_i^{(0)}, x_i^{(1)}, \dots, x_i^{(m)}) \in Z \qquad (i = 1, 2, \dots)$$

which satisfies, when $i \to \infty$,

1. $x_i \to 0$.
2. $x_i^{(k)}/|x_i^{(k)}| \to \xi^{(k)}/|\xi^{(k)}|$ for $k \notin J_Z(\xi)$.
3. For any $(f, v) \in F^q$,

$$f(|x_i^{(1)}|, \dots, |x_i^{(m)}|) \to v, \quad (i \to \infty).$$

By considering a subsequence, we may assume, for each $k \in \{1, 2, \dots, m\}$, either $x_i^{(k)} \neq 0$ for all i or $x_i^{(k)} = 0$ for all i. Note that latter case occurs only for $k \in J_Z(\xi)$.

Set $\tau_{k,i} = |x_i^{(k)}|$ for $1 \leq k \leq m$. Now we define $\tilde{\tau}_{k,i}$ as follows: For $k \in \{1, 2, \dots, m\}$ with $\tau_{k,i} \neq 0$ $(i = 1, 2, \dots)$, we set $\tilde{\tau}_{k,i} = \tau_{k,i}$. For $k \in \{1, 2, \dots, m\}$ with $\tau_{k,i}$ being identically zero $(i = 1, 2, \dots)$, we take, as a $\{\tilde{\tau}_{k,i}\}_{i=1}^{\infty}$, a sequence of positive real numbers satisfying $\tilde{\tau}_{k,i} \to 0$ $(i \to \infty)$ and

$$f(\tilde{\tau}_{1,i}, \dots, \tilde{\tau}_{m,i}) \to v, \quad (i \to \infty) \tag{2.48}$$

for any $(f, v) \in F^q$. Note that it is possible because such a k belongs to $J_Z(\xi)$ and we have $\nu_k(f) \geq 0$ for $k \in J_Z(\xi)$. Furthermore, if $\nu_k(f) > 0$ for some $k \in J_Z(\xi)$, then the corresponding v must be zero by Lemma 2.25. Therefore it suffices to take $\tilde{\tau}_{k,i}$ rapidly decreasing to 0 $(i \to \infty)$.

Recall that $J_Z(\xi) \cap \{1, 2, \dots, L\} = \{k_1, \dots, k_q\}$ $(1 \leq k_1 < \dots < k_q \leq L)$. Then, as F^q is generated by applying the operation $\tilde{\mathcal{L}}_{k_q}$ to F^{q-1}, by the conditions (2.48) and Lemma 2.33, we find a sequence $\{\epsilon_{k_q,i}\}$ of positive real numbers such that, when $i \to \infty$,

$$\epsilon_{k_q,i} \to 0, \quad f\left(\tilde{\tau}_{1,i}, \dots, \overset{k_q\text{-th}}{(\tilde{\tau}_{k_q,i}/\epsilon_{k_q,i})}, \dots, \tilde{\tau}_{m,i}\right) \to v \tag{2.49}$$

for any $(f, v) \in F^{q-1}$. Hence, thanks to the conditions (2.49), by applying the lemma again, we find a sequence $\{\epsilon_{k_{q-1},i}\}$ of positive real numbers such that $\epsilon_{k_{q-1},i} \to 0$ $(i \to \infty)$ and

$$f\left(\tilde{\tau}_{1,i}, \dots, \overset{k_{q-1}\text{-th}}{(\tilde{\tau}_{k_{q-1},i}/\epsilon_{k_{q-1},i})}, \dots, \overset{k_q\text{-th}}{(\tilde{\tau}_{k_q,i}/\epsilon_{k_q,i})}, \dots, \tilde{\tau}_{m,i}\right) \to v \quad (i \to \infty)$$

for any $(f, v) \in F^{q-2}$. By repeated applications of the lemma in this way, we finally find the q-sequences $\{\epsilon_{k,i}\}_{i=1}^{\infty}$ $(k \in J_Z(\xi) \cap \{1, 2, \dots, L\})$ of positive real numbers which satisfies, when $i \to \infty$,

$$\begin{aligned} &\epsilon_{k,i} \to 0 && (k \in J_Z(\xi) \cap \{1, 2, \dots, L\}), \\ &f(\hat{\tau}_{1,i}, \dots, \hat{\tau}_{m,i}) \to v && (f, v) \in F^0, \end{aligned}$$

where $\hat{\tau}_{k,i}$ denotes $\tilde{\tau}_{k,i}/\epsilon_{k,i}$ if $k \in J_Z(\xi) \cap \{1, 2, \dots, L\}$ and $\hat{\tau}_{k,i} = \tilde{\tau}_{k,i}$ for other k. Set

$$\hat{\tau}_i = (\hat{\tau}_{1,i}, \dots, \hat{\tau}_{m,i}) \text{ and } \hat{\tau}'_i = (\hat{\tau}_{1,i}, \dots, \hat{\tau}_{L,i}).$$

Now applying the same argument to the variables $\lambda'' = (\lambda_{L+1}, \dots, \lambda_\ell)$ repeatedly, we can find sequences

$$\lambda''_i = (\lambda_{L+1,i}, \dots, \lambda_{\ell,i}) \quad (i = 1, 2, \dots)$$

of positive real numbers such that, when $i \to \infty$,

$$\lambda''_i \to 0, \qquad f(\hat{\tau}_i, \lambda''_i) \to v$$

for any $(f, v) \in G$. Finally we define the sequence in $\widetilde{X}$ by

$$\begin{aligned}\widetilde{x}_i := \big(&x_i^{(0)}, (x_i^{(1)}/\hat{\tau}_{1,i}), \dots, (x_i^{(L)}/\hat{\tau}_{L,i}),\\ &(x_i^{(L+1)}/\varphi_{L+1}(t_i)), \dots, (x_i^{(m)}/\varphi_m(t_i)), t_{1,i}, \dots, t_{\ell,i}\big) \in \widetilde{X},\end{aligned}$$

where

$$t_{j,i} := \varphi_j^{-1}(\hat{\tau}'_i, \lambda''_i)^{1/\sigma_A} \qquad (1 \le j \le L),$$

$$t_{j,i} := \lambda_{j,i}^{1/\sigma_A} \qquad (L < j \le \ell).$$

Let $\hat{p} = (0; \hat{\xi}^{(1)}, \dots, \hat{\xi}^{(\ell)})$ be the normalized point of p. That is,

$$\hat{\xi}^{(k)} := \begin{cases} 0 & (k \in J_Z(\xi) \cap \{1, \dots, L\}),\\ \dfrac{\xi^{(k)}}{|\xi^{(k)}|} & (1 \le k \le L,\ \xi^{(k)} \ne 0),\\ \dfrac{n_k(\xi)\xi^{(k)}}{|\xi^{(k)}|} & (k > L).\end{cases}$$

Note that p and $\hat{p}$ belong to the same orbit of $(\mathbb{R}^+)^\ell$-actions on S_χ. Then we find $\widetilde{p}(\widetilde{x}_i) = x_i \in Z$ and $\widetilde{x}_i \to \hat{p}$ $(i \to \infty)$, which implies $\hat{p} \in C_\chi(Z)$. Since $C_\chi(Z)$ is an $(\mathbb{R}^+)^\ell$-conic subset, we have finally obtained $p \in C_\chi(Z)$. This completes the proof. □

By the same argument as that in the first part of the proof of the theorem, we see that a multi-cone enjoys the following good property.

Corollary 2.36. *Let H be a finite subset in $\mathcal{H}_\tau \times \mathbb{R}_{\ge 0}$ for which $[H]$ and $\mathcal{G}$ are equivalent. Then every $V \in C(\xi, H)$ is stable under contraction induced by the action μ_j $(j = 1, \dots, \ell)$. That is, $\mu_j(x, \lambda_j) \in V$ holds for any $x \in V$ and any $0 < \lambda_j \le 1$.*

Proof. Let $(f, v) \in H$ and $1 \le j_0 \le \ell$. For simplicity, we set $f(z) := f(|z^{(*)}|)$. To show the corollary, it suffices to prove the following claim:

1. If $v \neq 0$, then $f(\mu_{j_0}(z, \lambda_{j_0})) = f(z)$.
2. If $v = 0$, there exist a non-negative rational number κ such that $f(\mu_{j_0}(z, \lambda_{j_0})) = \lambda_{j_0}^{\kappa} f(z)$.

Since $[H]$ is equivalent to $\widehat{\mathcal{G}}$ also, by the definition, there exist $N \in \mathbb{Z}$, $\alpha_k \in \mathbb{Z}$ and $\beta_j \in \mathbb{Z}_{\geq 0}$ satisfying

$$f(\tau)^N = \prod_{1 \leq k \leq m} (\tau_k/\varphi(\lambda))^{\alpha_k} \prod_{1 \leq j \leq \ell} \lambda_j^{\beta_j}.$$

Put $\tau = |(\mu_{j_0}(z, \lambda_{j_0}))^{(*)}|$ and $\lambda = (1, \ldots, 1, \lambda_{j_0}, 1, \ldots, 1)$ into the above equation, we have

$$f(\mu_{j_0}(z, \lambda_{j_0}))^N = \lambda_{j_0}^{\beta_{j_0}} f(z)^N.$$

By noticing that $v \neq 0$ implies $\beta_j = 0$ for any j, we have obtained the claim. □

Now assume that $p = (x^{(0)}; \xi) \in S_\chi$ is outside fixed points and that the action μ_χ is non-degenerate. Then, by Corollary 2.5, we can take coordinates blocks of a local model so that the conditions 1. and 2. of the corollary hold. Furthermore, under this coordinates blocks, since $L = \ell$ and $q = 0$ hold, we have $F^q = G$. Hence we have obtained the following corollary: Let us recall the definition of $n_k(\xi)$ given in Remark 2.10 (note that $n_k(\xi)$ is just $|\xi^{(k)}|$ when ξ is normalized).

Corollary 2.37. *Under the situation described above, the following family of subsets gives a cofinal family of $C(\xi)$:*

$$\left\{ (x^{(0)}, x^{(1)}, \ldots, x^{(m)});\ \begin{array}{ll} x^{(k)} \in W_k & (k = 1, \ldots, m), \\ \varphi_k^{-1}(|x^{(*)}|') < \epsilon & (k \leq \ell), \\ n_k(\xi) - \epsilon < \dfrac{|x^{(k)}|}{\varphi_k(\varphi^{-1}(|x^{(*)}|'))} < n_k(\xi) + \epsilon & (k > \ell) \end{array} \right\},$$

where $\epsilon > 0$, and W_k runs through open conic cones in $\mathbb{R}^{n_k} \setminus \{0\}$ containing the direction $\xi^{(k)}$ if $\xi^{(k)} \neq 0$ and it is $\mathbb{R}^{n_k}$ if $\xi^{(k)} = 0$.

Example 2.38. *Let $M_1 = \{x_1 = x_2 = 0\}$, $M_2 = \{x_2 = x_3 = 0\}$ and consider the action on X by*

$$\mu(x, t) = (x_1 t_1, x_2 t_1 t_2, x_3 t_2).$$

The associated matrix is

$$A_\chi = \begin{pmatrix} 1 & 1 & 0 \\ 0 & 1 & 1 \end{pmatrix}.$$

We have seen that $p = (\xi_1, \xi_2, \xi_3) \in S_\chi$ is outside the set of fixed points if two of the three coordinates are different from 0. If $\xi_1, \xi_2 \neq 0$ we set $\varphi_1(\lambda) = \lambda_1$, $\varphi_2(\lambda) = \lambda_1\lambda_2$, $\varphi_3(\lambda) = \lambda_2$. The invertible 2×2 matrix is

$$A'_\chi = \begin{pmatrix} 1 & 1 \\ 0 & 1 \end{pmatrix}, \quad A'^{-1}_\chi = \begin{pmatrix} 1 & -1 \\ 0 & 1 \end{pmatrix}.$$

We have $\varphi_1^{-1}(\tau) = \tau_1$, $\varphi_2^{-1}(\tau) = \dfrac{\tau_2}{\tau_1}$, $\psi_3(\tau) = \dfrac{\tau_3\tau_1}{\tau_2}$. *Then*

$$G = Q\left(\left\{(\tau_1, 0), \left(\frac{\tau_2}{\tau_1}, 0\right), \left(\frac{\tau_1\tau_3}{\tau_2}, |\xi_3|\right)\right\}\right).$$

A cofinal family of $C(\xi)$ *is given by*

$$\left\{ (x_1, x_2, x_3);\ \begin{array}{ll} x_k \in W_k & (k = 1, 2, 3), \\ |x_1| < \epsilon, & \\ |x_2| < \epsilon |x_1|, & \\ |\xi_3| - \epsilon < \dfrac{|x_1||x_3|}{|x_2|} < |\xi_3| + \epsilon & \end{array} \right\},$$

where $\epsilon > 0$, *and* W_k *runs through open conic cones in* $\mathbb{R} \setminus \{0\}$ *containing the direction* ξ_k *if* $\xi_k \neq 0$ *and it is* $\mathbb{R}$ *if* $\xi_k = 0$.

Example 2.39. *Let* $M_1 = \{x_1 = x_3 = 0\}$, $M_2 = \{x_2 = x_3 = 0\}$ *and consider the action on* X *by*

$$\mu(x, t) = (x_1 t_1, x_2 t_2, x_3 t_1 t_2).$$

The associated matrix is

$$A_\chi = \begin{pmatrix} 1 & 0 & 1 \\ 0 & 1 & 1 \end{pmatrix}.$$

We have seen that $p = (\xi_1, \xi_2, \xi_3) \in S_\chi$ *is outside the set of fixed points if two of the three coordinates are different from* 0. *If* $\xi_1, \xi_2 \neq 0$ *we set* $\varphi_1(\lambda) = \lambda_1$, $\varphi_2(\lambda) = \lambda_2$, $\varphi_3(\lambda) = \lambda_1\lambda_2$. *The invertible* 2×2 *matrix is*

$$A'_\chi = \begin{pmatrix} 1 & 0 \\ 0 & 1 \end{pmatrix}, \quad A'^{\,-1}_\chi = \begin{pmatrix} 1 & 0 \\ 0 & 1 \end{pmatrix}.$$

We have $\varphi_1^{-1}(\tau) = \tau_1$, $\varphi_2^{-1}(\tau) = \tau_2$, $\psi_3(\tau) = \dfrac{\tau_3}{\tau_1\tau_2}$. *Then*

$$G = Q\left(\left\{(\tau_1, 0), (\tau_2, 0), \left(\frac{\tau_3}{\tau_1\tau_2}, |\xi_3|\right)\right\}\right).$$

A cofinal family of $C(\xi)$ *is given by*

$$\left\{ (x_1, x_2, x_3);\ \begin{array}{ll} x_k \in W_k & (k = 1, 2, 3), \\ |x_1| < \epsilon, & \\ |x_2| < \epsilon, & \\ |\xi_3| - \epsilon < \dfrac{|x_3|}{|x_1||x_2|} < |\xi_3| + \epsilon & \end{array} \right\},$$

where $\epsilon > 0$, *and* W_k *runs through open conic cones in* $\mathbb{R} \setminus \{0\}$ *containing the direction* ξ_k *if* $\xi_k \neq 0$ *and it is* $\mathbb{R}$ *if* $\xi_k = 0$.

Example 2.40. *Let $M_1 = \{x_1 = x_3 = 0\}$, $M_2 = \{x_1 = x_2 = 0\}$ and consider the action on X by*

$$\mu(x,t) = (x_1t_1t_2, x_2t_2, x_3t_1).$$

The associated matrix is

$$A_X = \begin{pmatrix} 1 & 0 & 1 \\ 1 & 1 & 0 \end{pmatrix}.$$

We have seen that $p = (\xi_1, \xi_2, \xi_3) \in S_X$ is outside the set of fixed points if two of the three coordinates are different from 0. If $\xi_1, \xi_2 \neq 0$ we set $\varphi_1(\lambda) = \lambda_1\lambda_2$, $\varphi_2(\lambda) = \lambda_2$, $\varphi_3(\lambda) = \lambda_1$. The invertible 2×2 matrix is

$$A'_X = \begin{pmatrix} 1 & 0 \\ 1 & 1 \end{pmatrix}, \quad A'^{-1}_X = \begin{pmatrix} 1 & 0 \\ -1 & 1 \end{pmatrix}.$$

We have $\varphi_1^{-1}(\tau) = \dfrac{\tau_1}{\tau_2}$, $\varphi_2^{-1}(\tau) = \tau_2$, $\psi_3(\tau) = \dfrac{\tau_2\tau_3}{\tau_1}$. Then

$$G = Q\left(\left\{\left(\frac{\tau_1}{\tau_2}, 0\right), (\tau_2, 0), \left(\frac{\tau_2\tau_3}{\tau_1}, |\xi_3|\right)\right\}\right).$$

A cofinal family of $C(\xi)$ is given by

$$\left\{ (x_1, x_2, x_3);\ \begin{array}{l} x_k \in W_k \qquad (k = 1,2,3), \\ |x_1| < \epsilon |x_2|, \\ |x_2| < \epsilon, \\ |\xi_3| - \epsilon < \dfrac{|x_2||x_3|}{|x_1|} < |\xi_3| + \epsilon \end{array} \right\},$$

where $\epsilon > 0$, and W_k runs through open conic cones in $\mathbb{R} \setminus \{0\}$ containing the direction ξ_k if $\xi_k \neq 0$ and it is $\mathbb{R}$ if $\xi_k = 0$.

2.4 A restriction condition

Let $A = (\alpha_{i,j})_{\substack{1 \le i \le \ell \\ 1 \le j \le m}}$ be an $\ell \times m$ matrix with entries of non-negative rationals. Set $L = \operatorname{rank} A$ $(1 \le L \le \min\{\ell, m\})$. Assume that the first $L \times L$ sub-matrix A' of A is invertible. Hereafter α_i denotes the i-th row of A for $1 \le i \le \ell$.

Lemma 2.41. *For $1 \le i \le \ell$, set $e^{\alpha_i} = (e^{\alpha_{i,1}}, \dots, e^{\alpha_{i,m}})$.*

(i) Let $1 \le i \le L$. We have $\log \varphi_j^{-1}(e^{\alpha_i})|_{\lambda''=1} \ge 0$ for $j = 1, \dots, L$.

(ii) Let $1 \le i \le \ell$. We have $\log \psi_k(e^{\alpha_i}) = 0$ for $k = L+1, \dots, m$.

Proof. Let $\beta_j = (\beta_{j,1}, \dots, \beta_{j,L})$, where $(\beta_{j,1}, \dots, \beta_{j,m})$ is the j-th column of A'^{-1} and $\beta_{j,k} = 0$, $k = L+1, \dots, m$.

(i) Let $\varphi_j^{-1}(\tau')$, $j = 1, \dots, L$. Then $\log \varphi_j^{-1}(e^{\alpha_i})|_{\lambda''=1}$ is nothing but the scalar product $\alpha_i \cdot \beta_j = \delta_{ij}$, where δ_{ij} is the Kronecker's delta.

(ii) First suppose that $1 \leq i \leq L$. Let $\psi_k(\tau')$, $k = L+1, \ldots, m$. Then $\log \psi_k(e^{\alpha_i})$ is nothing but the scalar product $\alpha_i \cdot \left(e_k - \sum_{j=1}^m \alpha_{jk}\beta_j\right) = \alpha_{ik} - \alpha_{ik} = 0$. Moreover $\log(1/\psi_k(e^{\alpha_i})) = 0$ as well.
Now suppose that $L+1 \leq i \leq \ell$. Then $\alpha_i = \sum_{j=1}^L c_j\alpha_j$, so $\log \psi_k(e^{\alpha_i}) = \sum_{j=1}^L c_j \log \psi_k(e^{\alpha_j}) = 0$. Moreover $\log(1/\psi_k(e^{\alpha_i})) = 0$ as well.

□

Corollary 2.42. *Let $(f, v) \in \mathcal{G}$ and $1 \leq j \leq L$. Then*

$$\begin{aligned} &\log f(e^{\alpha_j}) = 0 \quad \text{if } f = \textstyle\prod_{k=L+1}^m \psi_k, \\ &\log f(e^{\alpha_j}) \geq 0 \quad \text{otherwise.} \end{aligned} \tag{2.50}$$

In particular

$$\begin{aligned} v = 0 &\Rightarrow \log f(e^{\alpha_j}) \geq 0, \\ v \neq 0 &\Rightarrow \log f(e^{\alpha_j}) = 0. \end{aligned} \tag{2.51}$$

Proof. This follows immediately from Lemma 2.41 and the fact that $[G]_=$ and $[G]_> * (1,0)$ generate $\mathcal{G}$.

□

Let us consider a $\ell \times m$ matrix A and a $(\ell+1) \times m$ matrix B obtained by adding to A a $(\ell+1)$-th row $(\beta_1, \ldots, \beta_m)$ ($\beta_i \in \mathbb{Q}_{\geq 0}$). We will use $(\cdot)_A$ (resp. $(\cdot)_B$) to indicate elements (sets, family of sets) related to the normal deformation associated with A (resp. B). There are two possibilities: $\operatorname{rank} B = L$ or $\operatorname{rank} B = L+1$.

Let us consider the case $\operatorname{rank} B = L$. Set

$$b_j := \log \varphi_{jA}^{-1}(e^\beta)|_{\lambda''=1} \qquad (j = 1, \ldots, L). \tag{2.52}$$

Lemma 2.43. *Let us assume that*

$$v = 0 \Rightarrow \log f(e^\beta) \geq 0 \tag{2.53}$$

for $(f, v) \in \mathcal{G}_A$. Then $\mathcal{G}_A$ and $\mathcal{G}_B$ have the same radical.

Proof. For $j = 1, \ldots, L$, we have

$$\tau_j = \varphi_{jB}(\lambda', \lambda'', \lambda_{\ell+1}) = \varphi_{jA}(\lambda', \lambda'') \cdot \lambda_{\ell+1}^{\beta_j}.$$

After a simple computation we obtain

$$\lambda_j = \varphi_{jA}^{-1}(\tau, \lambda'')\lambda_{\ell+1}^{-b_j} = \varphi_{jB}^{-1}(\tau, \lambda'', \lambda_{\ell+1}).$$

Hence, for $k \geq L+1$, we have

$$\psi_{kB}(\tau) = \psi_{kA}(\tau).$$

Set $J = \{1, \ldots, L\}$, $K = \{L+1, \ldots, m\}$ and $\tilde{J} = \{L+1, \ldots, \ell\}$. If $v \neq 0$, then $f = \prod_{k \in K} \psi_k^{\delta_k}$ for some $\delta_k \in \mathbb{Z}$, and hence, $(f, v) \in \mathcal{G}_A$ if and only if $(f, v) \in \mathcal{G}_B$.

Assume now $v = 0$. Let $f = \prod_{j \in J} (\varphi_{jA}^{-1})^{a_j} \cdot \prod_{k \in K} (\psi_{kA})^{a_k}$. Then

$$\begin{aligned} \log f(e^\beta)|_{\lambda''=1} &= \sum_{j \in J} a_j \log \varphi_{jA}^{-1}(e^\beta)|_{\lambda''=1} + \sum_{k \in K} a_k \log \psi_{kA}(e^\beta) \\ &= \sum_{j \in J} a_j \log \varphi_{jA}^{-1}(e^\beta)|_{\lambda''=1} + \sum_{k \in K} a_k \log \psi_{kB}(e^\beta) \\ &= \sum_{j \in J} a_j b_j, \end{aligned}$$

where the last equality follows since $\log \psi_{kB}(e^\beta) = 0$ by Lemma 2.41. Here we used the fact that $\operatorname{rank} A = \operatorname{rank} B$.

Now assume that $(f, 0) \in \mathcal{G}_A$. Set $b_f = \log f(e^\beta)|_{\lambda''=1}$. Then

$$\begin{aligned} f &= \prod_{j \in J} (\varphi_{jA}^{-1})^{a_j} \cdot \prod_{k \in K} (\psi_{kA})^{a_k} \cdot \prod_{j \in \tilde{J}} \lambda_j^{\delta_j} \\ &= \prod_{j \in J} (\varphi_{jB}^{-1})^{a_j} \cdot \lambda_{\ell+1}^{b_f} \cdot \prod_{k \in K} (\psi_{kB})^{a_k} \cdot \prod_{j \in \tilde{J}} \lambda_j^{\delta_j}, \end{aligned}$$

and $(f, 0) \in \mathcal{G}_B$ if $b_f \geq 0$. Conversely, assume that $(f, 0) \in \mathcal{G}_B$. Then

$$\begin{aligned} f &= \prod_{j \in J} (\varphi_{jB}^{-1})^{a_j} \cdot \lambda_{\ell+1}^{b_f} \cdot \prod_{k \in K} (\psi_{kB})^{a_k} \cdot \prod_{j \in \tilde{J}} \lambda_j^{\delta_j} \\ &= \prod_{j \in J} (\varphi_{jA}^{-1})^{a_j} \cdot \prod_{k \in K} (\psi_{kA})^{a_k} \cdot \prod_{j \in \tilde{J}} \lambda_j^{\delta_j}, \end{aligned}$$

where the term $\lambda_{\ell+1}^{b_f}$ appears since the degree in $\lambda_{\ell+1}$ must be 0. Then $(f, 0) \in \mathcal{G}_A$ and the result follows. □

Corollary 2.44. *Suppose that* $\beta = \sum_{j=1}^{L} c_j \alpha_j$ *with* $c_j \geq 0$, $j = 1, \ldots, L$. *Then* $\mathcal{G}_A$ *and* $\mathcal{G}_B$ *have the same radical.*

Proof. By Lemma 2.41 we obtain $b_f := \log f(e^\beta) \geq 0$ for each $(f, v) \in \mathcal{G}_A$ and the result follows by Lemma 2.43. □

Thanks to Proposition 2.28 it is enough to check the conditions on the elements of F_A^q, $q := \#(J_Z(\xi) \cap \{1, 2, \ldots, L\}) \geq 0$, and we have

Proposition 2.45. *Let us assume that* $\operatorname{rank} B = \operatorname{rank} A$ *and*

$$v = 0 \Rightarrow \log f(e^\beta) \geq 0 \tag{2.54}$$

for $(f, v) \in F_A^q$. *Then* $\mathcal{G}_A$ *and* $\mathcal{G}_B$ *have the same radical.*

From Proposition 2.45 we can deduce a condition for the restriction of multi-normal deformations. We omit the coordinate $x^{(0)}$ to lighten notations. In what follows, given $\xi = (\xi^{(1)}, \dots, \xi^{(m)}) \in \mathbb{R}^n$ we will call $p_A \in S_A$, the point $p_A = (\xi^{(1)}, \dots, \xi^{(m)}; 0 \dots, 0) \in \mathbb{R}^{n+\ell}$, $p_B \in S_B$, the point $p_B = (\xi^{(1)}, \dots, \xi^{(m)}; 0 \dots, 0) \in \mathbb{R}^{n+\ell+1}$.

Proposition 2.46. *Let* $p_A = (\xi^{(1)}, \dots, \xi^{(m)}; 0, \dots, 0) \in S_A$. *Assume that* $\operatorname{rank} A = \operatorname{rank} B$ *and*

$$v = 0 \Rightarrow \log f(e^{\beta}) \geq 0 \tag{2.55}$$

for $(f, v) \in F_A^q$. *Then* $p_B \in C_{\chi_B}(Z)$ *if and only if, for each* $V \in C(\xi, \mathcal{G}_A)$ *and each open neighborhood* U *of the origin, we have* $Z \cap V \cap U \neq \emptyset$ *(i.e., if and only if* $p_A \in C_{\chi_A}(Z)$*).*

Thanks to Corollary 2.44 we obtain a simpler sufficient condition. Recall $L := \operatorname{rank} A$.

Corollary 2.47. *Let* $p_A = (\xi^{(1)}, \dots, \xi^{(m)}; 0, \dots, 0) \in S_A$ *and assume that* $\beta = \sum_{j=1}^{L} c_j \alpha_j$ *with* $c_j \geq 0$, $j = 1, \dots, L$. *Then* $p_B \in C_{\chi_B}(Z)$ *if and only if* $p_A \in C_{\chi_A}(Z)$.

Example 2.48. *In* $\mathbb{R}^3$ *let us consider the families* $\chi_A = \{M_1, M_2, M_3\}$, $\chi_B = \{M_1, M_2, M_3, M_4\}$ *with* $M_1 = \{x_1 = 0\}$, $M_2 = \{x_2 = x_3 = 0\}$, $M_3 = \{x_3 = 0\}$, $M_4 = \{0\}$. *If we consider the matrices*

$$A = \begin{pmatrix} 1 & 0 & 0 \\ 0 & 1 & 1 \\ 0 & 0 & 1 \end{pmatrix}, \quad B = \begin{pmatrix} 1 & 0 & 0 \\ 0 & 1 & 1 \\ 0 & 0 & 1 \\ 1 & 1 & 1 \end{pmatrix},$$

the hypothesis of Corollary 2.47 is satisfied.

In the following example, since v of all the pairs $(f, v) \in F^q$ are zero, we simply write a pair by f instead of (f, v).

Example 2.49. *In* $\mathbb{R}^3$ *let us consider the families* $\chi_A = \{M_1, M_2, M_3\}$, $\chi_B = \{M_1, M_2, M_3, M_4\}$ *with* $M_1 = \{x_1 = x_3 = 0\}$, $M_2 = \{x_2 = x_3 = 0\}$, $M_3 = \{x_3 = 0\}$, $M_4 = \{0\}$. *Set* $\beta = (1, 1, 1)$. *Consider the matrices*

$$A = \begin{pmatrix} 1 & 0 & 1 \\ 0 & 1 & 1 \\ 0 & 0 & 1 \end{pmatrix}, \quad B = \begin{pmatrix} 1 & 0 & 1 \\ 0 & 1 & 1 \\ 0 & 0 & 1 \\ 1 & 1 & 1 \end{pmatrix}.$$

1. *Let* $\xi_1, \xi_2, \xi_3 \neq 0$. *Then* $F_A^0 = \left\{ \tau_1, \tau_2, \dfrac{\tau_3}{\tau_1 \tau_2} \right\}$. *We have* $\log e^{\beta_1 - \beta_2 - \beta_3} = 1 - 1 - 1 < 0$ *and* (2.54) *is not satisfied. Indeed one can check that* $F_B^0 = \left\{ \tau_1, \tau_2, \dfrac{\tau_3}{\tau_1}, \dfrac{\tau_3}{\tau_2} \right\}$ *and* $\dfrac{\tau_3}{\tau_1 \tau_2} \notin \mathcal{G}_B$.

2. *Let* $\xi_3 = 0$, $\xi_1, \xi_2 \neq 0$. *Then* $F_A^1 = \left\{\tau_1, \tau_2, \dfrac{\tau_3}{\tau_1\tau_2}\right\}$. *We have* $\log e^{\beta_1-\beta_2-\beta_3} = 1-1-1 < 0$ *and* (2.54) *is not satisfied. Indeed one can check that* $F_B^1 = \left\{\tau_1, \tau_2, \dfrac{\tau_3}{\tau_1}, \dfrac{\tau_3}{\tau_2}\right\}$ *and* $\dfrac{\tau_3}{\tau_1\tau_2} \notin \mathcal{G}_B$.

3. *Let* $\xi_2 = 0$, $\xi_1, \xi_3 \neq 0$. *Then* $F_A^1 = \left\{\tau_1, \tau_2, \dfrac{\tau_3}{\tau_1}\right\}$ *and condition* (2.54) *is satisfied.*

4. *Let* $\xi_1 = 0$, $\xi_2, \xi_3 \neq 0$. *Then* $F_A^1 = \left\{\tau_1, \tau_2, \dfrac{\tau_3}{\tau_2}\right\}$ *and condition* (2.54) *is satisfied.*

Let us consider the case $\operatorname{rank} B = L+1$. If $L < \ell$, we exchange the $(L+1)$-th line with the $(\ell+1)$-th line.

Let $\varphi_{jA}^{-1}(\tau, \lambda'')$, $j = 1, \ldots, L$ and $\psi_{kA}(\tau)$, $k = L+1, \ldots, m$ be the monomials defining $\mathcal{G}_A$. We are going to compute $\varphi_{jB}^{-1}(\tau, \lambda'')$, $j = 1, \ldots, L+1$ and $\psi_{kB}(\tau)$, $k = L+2, \ldots, m$ using $\varphi_{jA}^{-1}(\tau, \lambda'')$ and $\psi_{kA}(\tau)$. Set

$$b_j = \begin{cases} \log \varphi_{jA}^{-1}(e^{\beta})|_{\lambda''=1} & (j = 1, \ldots, L), \\ \log \psi_{jA}(e^{\beta}) & (j = L+1, \ldots, m). \end{cases}$$

Assume for simplicity that all the columns of B are non zero. We may also assume that $\exists\, k \in \{L+1, \ldots, m\}$ such that $b_k \neq 0$. Otherwise $\log \psi_{kA}(e^{\alpha_j}) = 0$ for $j = 1, \ldots, L+1$ by Lemma 2.41. This implies that the rank of B is the same as the rank of A. So, up to take a permutation of $\{1, \ldots, m\}$, we may assume that $b_{L+1} \neq 0$. If $b_{L+1} \neq 0$, then the first $(L+1) \times (L+1)$ sub-matrix B' of B is invertible. Otherwise $\operatorname{rank} B' = \operatorname{rank} A'$ and, as in the proof of Lemma 2.43, $\log \psi_{(L+1)B'}(e^{\beta}) = \log \psi_{(L+1)A'}(e^{\beta})$. We have

$$\log \psi_{(L+1)B'}(e^{\beta}) = \log \psi_{(L+1)A'}(e^{\beta}) = \log \psi_{(L+1)A}(e^{\beta}) = b_{L+1}$$

and $\log \psi_{(L+1)B'}(e^{\beta}) = 0$ by Lemma 2.41.

Let τ denote m-variables $(\tau_1, \ldots, \tau_m)$. Consider the $(L+1)$-equations

$$\tau_1 = \varphi_{1B}(\lambda), \ldots, \tau_{L+1} = \varphi_{L+1B}(\lambda).$$

It can be also written as

$$\tau_1 = \varphi_{1A}(\lambda)\lambda_{L+1}^{\beta_1}, \ldots, \tau_{L+1} = \varphi_{L+1A}(\lambda)\lambda_{L+1}^{\beta_{L+1}}.$$

Then, by the assumption, the system of the equations can be solved for the variables

$\lambda = (\lambda_1, \ldots, \lambda_{L+1})$. After a computation, we can find

$$\begin{aligned}
\varphi_{jB}^{-1}(\tau, \lambda'') &= \frac{\varphi_{jA}^{-1}(\tau, \lambda'')}{(\psi_{L+1A}(\tau))^{\frac{b_j}{b_{L+1}}}} \quad (j = 1, \ldots, L), \\
\varphi_{L+1B}^{-1}(\tau) &= (\psi_{L+1A}(\tau))^{\frac{1}{b_{L+1}}}, \\
\psi_{kB}(\tau) &= \frac{\psi_{kA}(\tau)}{(\psi_{L+1A}(\tau))^{\frac{b_k}{b_{L+1}}}} \quad (k = L+2, \ldots, m),
\end{aligned}$$

and

$$\begin{aligned}
\varphi_{jA}^{-1}(\tau, \lambda'') &= \varphi_{jB}^{-1}(\tau, \lambda'') \cdot (\varphi_{L+1B}^{-1}(\tau))^{b_j} \quad (j = 1, \ldots, L), \\
\psi_{L+1A}(\tau) &= (\varphi_{L+1B}^{-1}(\tau))^{b_{L+1}}, \\
\psi_{kA}(\tau) &= \psi_{kB}(\tau) \cdot (\varphi_{L+1B}^{-1}(\tau))^{b_k} \quad (k = L+2, \ldots, m).
\end{aligned}$$

In what follows, we omit the variables τ, λ'' to lighten notations.

Lemma 2.50. *Let us assume that*

$$\begin{aligned}
v = 0 &\Rightarrow \log f(e^\beta) \geq 0, \\
v \neq 0 &\Rightarrow \log f(e^\beta) = 0, \\
\text{if } m \geq L+2, \quad b_k &= 0 \quad \forall k \in \{L+2, \ldots, m\} \setminus J_Z(\xi)
\end{aligned} \tag{2.56}$$

for $(f, v) \in \mathcal{G}_A$. Then $\mathcal{G}_A$ and $\mathcal{G}_B$ have the same radical.

Proof. First remark that $b_{L+1} \neq 0$ and (2.56) imply $((\psi_{L+1A})^{a_{L+1}}, v) \in \mathcal{G}_A$, $a_{L+1} \neq 0$, only if $v = 0$. Since the variable λ'' does not play a role in this proof, we assume $\ell = L$.

(i) Let $(f_A, v) \in \mathcal{G}_A$. Then

$$\begin{aligned}
f_A &= \prod_{j \in J} (\varphi_{jA}^{-1})^{a_j} \cdot (\psi_{\ell+1A})^{a_{\ell+1}} \cdot \prod_{k \in K} (\psi_{kA})^{a_k} \\
&= \prod_{j \in J} (\varphi_{jB}^{-1})^{a_j} \cdot (\varphi_{\ell+1B}^{-1})^{c} \cdot \prod_{k \in K} (\psi_{kB})^{a_k},
\end{aligned}$$

$J \subseteq \{1, \ldots, \ell\}$, $K \subseteq \{\ell+2, \ldots, m\}$ and

$$c = \sum_{j \in J} a_j b_j + a_{\ell+1} b_{\ell+1} + \sum_{k \in K} a_k b_k = \log f_A(e^\beta).$$

- Suppose that $v = 0$. It follows from (2.56) that $c \geq 0$.
  - Suppose that $c > 0$. Then $(f_A, 0) \in \mathcal{G}_B$ since $((\varphi_{\ell+1B}^{-1})^c, 0) \in [G_B]$ and $(\varphi_{\ell+1B}^{-1}, v) \in G_B$ only if $v = 0$.

– Suppose that $c = 0$. Then there are two possibilities:

  * $a_j \neq 0$, $\exists j \in J$ or $a_k \neq 0$, $\exists k \in K \cap J_Z(\xi)$. In this case $(f_A, 0) \in \mathcal{G}_B$ since $((\varphi_{jB}^{-1})^{a_j}, 0), ((\psi_{kB})^{a_k}, 0) \in [G_B]$ and $(\varphi_{jB}^{-1}, v), (\psi_{kB}, v) \in G_B$ only if $v = 0$.

  * $a_j = 0$, $\forall j \in J$ and $a_k = 0$, $\forall k \in K \cap J_Z(\xi)$. We may assume $K \cap J_Z(\xi) = \emptyset$ and by (2.56) we have

$$f_A = \prod_{k \in K} \frac{(\psi_{kA})^{a_k}}{(\psi_{\ell+1A})^{\frac{a_k b_k}{b_{\ell+1}}}} = \prod_{k \in K} (\psi_{kA})^{a_k} = \prod_{k \in K} (\psi_{kB})^{a_k}.$$

  This is because by (2.56) we have $b_k = 0$, $k \in K$. We have $(f_A, v) \in [G_A]_> * [(1,0)]$. Hence $(f_A, 0) \in [G_B]_> * [(1,0)]$.

• Suppose that $v \neq 0$. Then $(f_A, v) \in [G_A]_=$. It follows from (2.56) that $c = 0$. Moreover $a_j = 0$, $j \in J$ and $a_k = 0$, $k \in K \cap J_Z(\xi)$ since $(\varphi_{jA}^{-1}, v), (\psi_{kA}, v) \in G_A$ only if $v = 0$. Then

$$f_A = \prod_{k \in K \setminus J_Z(\xi)} \frac{(\psi_{kA})^{a_k}}{(\psi_{\ell+1A})^{\frac{a_k b_k}{b_{\ell+1}}}} = \prod_{k \in K \setminus J_Z(\xi)} (\psi_{kA})^{a_k} = \prod_{k \in K \setminus J_Z(\xi)} (\psi_{kB})^{a_k}.$$

This is because by (2.56) we have $b_k = 0$ if $k \notin J_Z(\xi)$. Then $(f_A, v) \in [G_A]_=$ implies $(f_A, v) \in [G_B]_=$.

(ii) Let $(f_B, v) \in \mathcal{G}_B$. Then

$$\begin{aligned} f_B &= \prod_{j \in J} (\varphi_{jB}^{-1})^{a_j} \cdot (\varphi_{\ell+1B}^{-1})^{a_{\ell+1}} \cdot \prod_{k \in K} (\psi_{kB})^{a_k} \\ &= \prod_{j \in J} (\varphi_{jA}^{-1})^{a_j} \cdot (\psi_{\ell+1A})^{c} \cdot \prod_{k \in K} (\psi_{kA})^{a_k}, \end{aligned}$$

$J \subseteq \{1, \ldots, \ell\}$, $K \subseteq \{\ell+2, \ldots, m\}$ and

$$c = \frac{a_{\ell+1} - \sum_{j \in J} a_j b_j - \sum_{k \in K} a_k b_k}{b_{\ell+1}}.$$

• Suppose that $v = 0$. Then $((f_B)^{|b_{\ell+1}|}, w) \in \mathcal{G}_A$ for some $w \geq 0$. In fact, it clearly holds if $\xi^{(\ell+1)} \neq 0$. Otherwise, as $(f_B, v) \in \mathcal{G}_B$ implies $\nu_{\ell+1}(f_B) \geq 0$ when $\xi^{(\ell+1)} = 0$, we conclude $c \geq 0$ since we see $c = \nu_{\ell+1}(f_B)$ by comparing the degree of the variable $\tau_{\ell+1}$.

  – Suppose $a_j \neq 0$, $\exists j \in J$ or $a_k \neq 0$, $\exists k \in K \cap J_Z(\xi)$. In this case we have $((f_B)^{|b_{\ell+1}|}, 0) \in \mathcal{G}_A$ since $((\varphi_{jA}^{-1})^{a_j}, 0), ((\psi_{kA})^{a_k}, 0) \in [G_A]$ and $(\varphi_{jA}^{-1}, v), (\psi_{kA}, v) \in G_A$ only if $v = 0$.

  – Suppose that $a_j = 0$, $\forall j \in J$ and $a_k = 0$, $\forall k \in K \cap J_Z(\xi)$. We may assume $K \cap J_Z(\xi) = \emptyset$. Let us argue by contradiction and suppose that $w \neq 0$. Then $((f_B)^{|b_{\ell+1}|}, w) \in [G_A]_=$. Corollary 2.42 implies that $a_{\ell+1} = \log f_B(e^\beta) = 0$ and by (2.56) we have

$$f_B = \prod_{k\in K} (\psi_{kB})^{a_k} = \prod_{k\in K} \frac{(\psi_{kA})^{a_k}}{(\psi_{\ell+1A})^{\frac{a_k b_k}{b_{\ell+1}}}} = \prod_{k\in K} (\psi_{kA})^{a_k}.$$

This implies that $((f_B)^{|b_{\ell+1}|}, w) \in [G_B]_=$, a contradiction. Hence we have $w = 0$ and $((f_B)^{|b_{\ell+1}|}, 0) \in \mathcal{G}_A$ follows.

- Suppose that $v \neq 0$. Then $(f_B, v) \in [G_B]_=$. In this case $a_j = 0$, $j \in J \cup \{\ell+1\}$ and $a_k = 0$, $k \in K \cap J_Z(\xi)$ since $(\varphi_{jB}^{-1}, v), (\psi_{kB}, v) \in G_B$ only if $v = 0$. Then

$$f_B = \prod_{k\in K\setminus J_Z(\xi)} (\psi_{kB})^{a_k} = \prod_{k\in K\setminus J_Z(\xi)} \frac{(\psi_{kA})^{a_k}}{(\psi_{\ell+1A})^{\frac{a_k b_k}{b_{\ell+1}}}} = \prod_{k\in K\setminus J_Z(\xi)} (\psi_{kA})^{a_k}.$$

This is because by (2.56) we have $b_k = 0$ if $k \notin J_Z(\xi)$. Then $(f_B, v) \in [G_B]_=$ implies $(f_B, v) \in [G_A]_=$.

□

Thanks to Proposition 2.28 it is enough to check the conditions on the elements of F_A^q, $q := \#(J_Z(\xi) \cap \{1, 2, \ldots, L\}) \geq 0$, and we have

Proposition 2.51. *Let us assume that*

$$\begin{aligned} v = 0 &\Rightarrow \log f(e^\beta) \geq 0, \\ v \neq 0 &\Rightarrow \log f(e^\beta) = 0, \\ \text{if } m \geq L+2, \quad b_k = 0 \ \ &\forall k \in \{L+2, \ldots, m\} \setminus J_Z(\xi) \end{aligned} \tag{2.57}$$

for $(f, v) \in F_A^q$. *Then* $\mathcal{G}_A$ *and* $\mathcal{G}_B$ *have the same radical.*

From Proposition 2.51 we can deduce a condition for the restriction of multi-normal deformations. We omit the coordinate $x^{(0)}$ to lighten notations. In what follows, given $\xi = (\xi^{(1)}, \ldots, \xi^{(m)}) \in \mathbb{R}^n$ we will call $p_A \in S_A$, the point $p_A = (\xi^{(1)}, \ldots, \xi^{(m)}; 0 \ldots, 0) \in \mathbb{R}^{n+\ell}$, $p_B \in S_B$, the point $p_B = (\xi^{(1)}, \ldots, \xi^{(m)}; 0 \ldots, 0) \in \mathbb{R}^{n+\ell+1}$.

Proposition 2.52. *Let* $p_A = (\xi^{(1)}, \ldots, \xi^{(m)}; 0, \ldots, 0) \in S_A$. *Assume that* rank $A+1 =$ rank B *and*

$$\begin{aligned} v = 0 &\Rightarrow \log f(e^\beta) \geq 0, \\ v \neq 0 &\Rightarrow \log f(e^\beta) = 0, \\ \text{if } m \geq L+2, \quad b_k = 0 \ \ &\forall k \in \{L+2, \ldots, m\} \setminus J_Z(\xi) \end{aligned} \tag{2.58}$$

for $(f, v) \in F_A^q$. *Then* $p_B \in C_{\chi_B}(Z)$ *if and only if, for each* $V \in C(\xi, \mathcal{G}_A)$ *and each open neighborhood* U *of the origin, we have* $Z \cap V \cap U \neq \emptyset$ *(i.e., if and only if* $p_A \in C_{\chi_A}(Z)$*).*

When χ_B is of normal type, i.e., $m = \ell + 1 = L + 1$, Proposition 2.52 can be reformulated as follows.

Corollary 2.53. *Let $m = \ell+1$. Let $p_A = (\xi^{(1)}, \dots, \xi^{(\ell+1)}; 0, \dots, 0) \in S_A$. Assume that $m = \operatorname{rank} B = \operatorname{rank} A + 1$ and*

$$\begin{gathered} v = 0 \Rightarrow \log f(e^{\beta}) \geq 0, \\ ((\psi_{\ell+1A})^{a_{\ell+1}}, v) \in \mathcal{G}_A \Rightarrow v = 0 \end{gathered} \tag{2.59}$$

for $(f, v) \in F^q_A$, $a_{\ell+1} \neq 0$. Then $p_B \in C_{\chi_B}(Z)$ if and only if, for each $V \in C(\xi, \mathcal{G}_A)$ and each open neighborhood U of the origin, we have $Z \cap V \cap U \neq \emptyset$ (i.e., if and only if $p_A \in C_{\chi_A}(Z)$).

Corollary 2.53 is a generalization of Corollary 4.3 of [9]. Consider the family $\chi_B = \{M_1, \dots, M_{\ell+1}\}$ satisfying condition H2 in [9], i.e.

H2 either $I_j \subset I_k$, $I_k \subset I_j$ or $I_j \cap I_k = \emptyset$ holds for $j \neq k$.

Consider a sub-family $\chi_A := \{M_{j_1}, \dots, M_{j_k}\}$ of χ_B. We denote by S_{χ_B} the zero section $\underset{X, 1 \leq j \leq \ell+1}{\times} T_{M_j} \iota_{\chi_B}(M_j)$ for the family χ_B. We also introduce the set

$$S_{\chi_B/\chi_A} := \left(\underset{X, 1 \leq \alpha \leq k}{\times} T_{M_{j_\alpha}} \iota_{\chi_B}(M_{j_\alpha}) \right) \underset{X}{\times} M, \tag{2.60}$$

where $M = \bigcap_{j=1}^{\ell+1} M_j$. We emphasize that, in the above definition, we use ι_{χ_B} (not ι_{χ_A}). Then we have the natural embedding

$$S_{\chi_B} \hookleftarrow S_{\chi_B/\chi_A} \hookrightarrow S_{\chi_A}.$$

Corollary 2.54. *Assume that χ_B satisfies the condition H2. Let $k \leq \ell$ and $\{j_1, \dots, j_k\}$ be a subset of $\{1, 2, \dots, \ell + 1\}$. Set $\chi_A = \{M_{j_1}, \dots, M_{j_k}\}$. Let Z be a subset of X. Then we have*

$$C_{\chi_B}(Z) \cap S_{\chi_B/\chi_A} = C_{\chi_A}(Z) \cap S_{\chi_B/\chi_A}.$$

Proof. It is enough to prove the result for $\sharp\chi_A = \ell$ and then apply it repeatedly. We can assume without loss of generality that $\chi_A = \{M_1, \dots, M_\ell\}$. We use the coordinates $x = (x^{(0)}, x^{(1)}, \dots, x^{(\ell+1)})$ defined as follows: $x^{(j)} = (x_i)_{i \in \widehat{I}_j}$ with

$$\begin{aligned} \widehat{I}_0 &= \{1, \dots, \ell + 1\} \setminus \bigcup_{i=1}^{\ell+1} I_i, \\ \widehat{I}_j &= I_j \setminus \bigcup_{I_i \subsetneq I_j} I_i, \quad j = 1, \dots, \ell + 1. \end{aligned}$$

Since the coordinate $x^{(0)}$ is irrelevant in the rest of the proof, we will assume $\widehat{I}_0 = \emptyset$. Then A is a $\ell \times (\ell + 1)$ matrix with entries $\alpha_{i,j}$, $1 \leq i \leq \ell$, $1 \leq j \leq \ell + 1$, B is a

$(\ell+1)\times(\ell+1)$ matrix obtained adding the line $(\alpha_{\ell+1,1},\dots,\alpha_{\ell+1,\ell+1})$. Up to take a permutation of the coordinates we may assume that the first ℓ columns of A are linearly independent. The entries of the matrices are given by $\alpha_{ij}=1$ if $\widehat{I_j}\subseteq I_i$, $\alpha_{ij}=0$ otherwise. The restriction maps $S_{\chi_B/\chi_A}\to S_{\chi_A}$, $S_{\chi_B/\chi_A}\to S_{\chi_B}$ are given by $\xi^{(\ell+1)}=0$.

Thanks to the condition H2 it is easy to compute the inverse matrix B^{-1}, whose columns define the monomials φ_{jB}^{-1}, $j=1,\dots,\ell+1$ which are given by

$$\begin{aligned}\varphi_{jB}^{-1}(\tau) &= \tau_j \text{ if } I_j \text{ is maximal},\\ \varphi_{jB}^{-1}(\tau) &= \frac{\tau_j}{\tau_{k_j}} \text{ otherwise, where } I_{k_j}=\bigcap_{I_k\supsetneqq I_j} I_k.\end{aligned}$$

Remark that to define φ_{jB}^{-1}, $j=1,\dots,\ell+1$ we are working with the family of indices I_j, $j=1,\dots,\ell+1$. Thanks to the condition H2 it is also easy to compute the inverse matrix A'^{-1} of A', the matrix defined by the first ℓ columns of A, whose columns define the monomials φ_{jA}^{-1}, $j=1,\dots,\ell$ which are given by

$$\begin{aligned}\varphi_{jA}^{-1}(\tau) &= \tau_j \text{ if } I_j \text{ is maximal},\\ \varphi_{jA}^{-1}(\tau) &= \frac{\tau_j}{\tau_{k_j}} \text{ otherwise, where } I_{k_j}=\bigcap_{I_k\supsetneqq I_j} I_k.\end{aligned}$$

Remark that to define φ_{jA}^{-1}, $j=1,\dots,\ell$ we are working with the family of indices I_j, $j=1,\dots,\ell$. Then one can also compute the monomial $\psi_{\ell+1A}$, which is given by

$$\begin{aligned}\psi_{\ell+1A}(\tau) &= \tau_{\ell+1} \text{ if } I_{\ell+1} \text{ is maximal},\\ \psi_{\ell+1A}(\tau) &= \frac{\tau_{\ell+1}}{\tau_{k_{\ell+1}}} \text{ otherwise, where } I_{k_{\ell+1}}=\bigcap_{I_k\supsetneqq I_{\ell+1}} I_k.\end{aligned}$$

It is easy to check that $\psi_{\ell+1A}(e^\beta)=1$ and $\varphi_{jA}^{-1}(e^\beta)\in\{0,1\}$. Hence the conditions (2.59) are satisfied and the result follows from Corollary 2.53. □

Example 2.55. *Let us consider some examples in $\mathbb{R}^3$ of Corollary 2.54.*

1. *(Majima) Let $X=\mathbb{R}^3$ with coordinates (x_1,x_2,x_3) and let $\chi_A=\{M_1,M_2\}$, $\chi_B=\{M_1,M_2,M_3\}$ with $M_i=\{x_i=0\}$, $i=1,2,3$. Then $F_A^0=F_B^0=Q(\{(\tau_1,0),(\tau_2,0),(\tau_3,|\xi_3|)\})$. If we consider $\xi_3=0$ and the matrices*

$$A=\begin{pmatrix}1&0&0\\0&1&0\end{pmatrix},\qquad B=\begin{pmatrix}1&0&0\\0&1&0\\0&0&1\end{pmatrix},$$

condition (2.57) is satisfied.

2. *(Takeuchi) Let* $X = \mathbb{R}^3$ *with coordinates* (x_1, x_2, x_3). *There are several configurations. In the subsequent examples, all the couples* (f, v) *are with* $v = 0$, *and we write it by* f *instead of* (f, v) *for short.*

   (a) *let* $\chi_A = \{M_1, M_2\}$, $\chi_B = \{M_1, M_2, M_3\}$ *with* $M_1 = \{0\}$, $M_2 = \{x_2 = x_3 = 0\}$, $M_3 = \{x_3 = 0\}$. *Let* $\xi_3 = 0$, $\xi_1, \xi_2 \neq 0$. *Then* $F_A^1 = \left\{\tau_1, \frac{\tau_2}{\tau_1}, \frac{\tau_3}{\tau_2}\right\}$. *If we consider the matrices*

$$A = \begin{pmatrix} 1 & 1 & 1 \\ 0 & 1 & 1 \end{pmatrix}, \qquad B = \begin{pmatrix} 1 & 1 & 1 \\ 0 & 1 & 1 \\ 0 & 0 & 1 \end{pmatrix},$$

   condition (2.57) *is satisfied.*

   (b) *let* $\chi_A = \{M_1, M_2\}$, $\chi_B = \{M_1, M_2, M_3\}$ *with* $M_1 = \{x_2 = x_3 = 0\}$, $M_2 = \{x_2 = 0\}$, $M_3 = \{0\}$. *Let* $\xi_3 = 0$, $\xi_1, \xi_2 \neq 0$. *Then* $F_A^1 = \left\{\tau_1, \frac{\tau_2}{\tau_1}, \tau_3\right\}$. *If we consider the matrices*

$$A = \begin{pmatrix} 1 & 1 & 0 \\ 0 & 1 & 0 \end{pmatrix}, \qquad B = \begin{pmatrix} 1 & 1 & 0 \\ 0 & 1 & 0 \\ 1 & 1 & 1 \end{pmatrix},$$

   condition (2.57) *is satisfied.*

   (c) *let* $\chi_A = \{M_1, M_2\}$, $\chi_B = \{M_1, M_2, M_3\}$ *with* $M_1 = \{0\}$, $M_2 = \{x_2 = 0\}$, $M_3 = \{x_2 = x_3 = 0\}$. *Let* $\xi_3 = 0$, $\xi_1, \xi_2 \neq 0$. *Then* $F_A^1 = \left\{\tau_1, \frac{\tau_2}{\tau_1}, \frac{\tau_3}{\tau_1}\right\}$. *If we consider the matrices*

$$A = \begin{pmatrix} 1 & 1 & 1 \\ 0 & 1 & 0 \end{pmatrix}, \qquad B = \begin{pmatrix} 1 & 1 & 1 \\ 0 & 1 & 0 \\ 0 & 1 & 1 \end{pmatrix},$$

   condition (2.57) *is satisfied.*

3. *(Mixed) Let* $X = \mathbb{R}^3$ *with coordinates* (x_1, x_2, x_3). *There are several configurations.*

   (a) *let* $\chi_A = \{M_1, M_2\}$, $\chi_B = \{M_1, M_2, M_3\}$ *with* $M_1 = \{0\}$, $M_2 = \{x_2 = 0\}$, $M_3 = \{x_3 = 0\}$. *Let* $\xi_3 = 0$, $\xi_1, \xi_2 \neq 0$. *Then* $F_A^1 = \left\{\tau_1, \frac{\tau_2}{\tau_1}, \frac{\tau_3}{\tau_1}\right\}$. *If we consider the matrices*

$$A = \begin{pmatrix} 1 & 1 & 1 \\ 0 & 1 & 0 \end{pmatrix}, \qquad B = \begin{pmatrix} 1 & 1 & 1 \\ 0 & 1 & 0 \\ 0 & 0 & 1 \end{pmatrix},$$

   condition (2.57) *is satisfied.*

(b) let $\chi_A = \{M_1, M_2\}$, $\chi_B = \{M_1, M_2, M_3\}$ with $M_1 = \{x_1 = 0\}$, $M_2 = \{x_2 = 0\}$, $M_3 = \{0\}$. Let $\xi_3 = 0$, $\xi_1, \xi_2 \neq 0$. Then $F_A^1 = \{\tau_1, \tau_2, \tau_3\}$. If we consider the matrices

$$A = \begin{pmatrix} 1 & 0 & 0 \\ 0 & 1 & 0 \end{pmatrix}, \qquad B = \begin{pmatrix} 1 & 0 & 0 \\ 0 & 1 & 0 \\ 1 & 1 & 1 \end{pmatrix},$$

condition (2.57) is satisfied.

Example 2.56. *In $\mathbb{R}^3$ let us consider the family $\chi_A = \{M_1, M_2\}$, with $M_1 = \{x_1 = x_2 = 0\}$, $M_2 = \{x_2 = x_3 = 0\}$. The matrix A is*

$$\begin{pmatrix} 1 & 1 & 0 \\ 0 & 1 & 1 \end{pmatrix}$$

and $\varphi_{1A}^{-1}(\tau) = \tau_1$, $\varphi_{2A}^{-1}(\tau) = \dfrac{\tau_2}{\tau_1}$, $\varphi_{3A}^{-1}(\tau) = \dfrac{\tau_1\tau_3}{\tau_2}$. We have

$$F_A^0 = \left\{(\tau_1, 0), \left(\frac{\tau_2}{\tau_1}, 0\right), \left(\frac{\tau_1\tau_3}{\tau_2}, |\xi_3|\right), \left(\frac{\tau_2}{\tau_1\tau_3}, \frac{1}{|\xi_3|}\right)\right\}.$$

In the following, all the couples (f, v) are with $v = 0$, so we omit v for short.

- *If $\xi_1 = 0$, $\xi_2, \xi_3 \neq 0$, $F_A^1 = \left\{\tau_1, \tau_2, \tau_3, \dfrac{\tau_1\tau_3}{\tau_2}, \dfrac{\tau_2}{\tau_3}\right\}$.*
- *If $\xi_2 = 0$, $\xi_1, \xi_3 \neq 0$, $F_A^1 = \left\{\tau_1, \dfrac{\tau_2}{\tau_1}, \tau_3, \dfrac{\tau_2}{\tau_1\tau_3}\right\}$.*
- *If $\xi_3 = 0$, $\xi_1, \xi_2 \neq 0$, $F_A^1 = \left\{\tau_1, \dfrac{\tau_2}{\tau_1}, \dfrac{\tau_1\tau_3}{\tau_2}\right\}$.*
- *If $\xi_2 \neq 0$, $\xi_1, \xi_3 = 0$, $F_A^2 = \left\{\tau_1, \tau_2, \tau_3, \dfrac{\tau_1\tau_3}{\tau_2}\right\}$.*

We study the restriction when we consider $\chi_B = \{M_1, M_2, M_3\}$ obtained by adding a manifold M_3 to χ_A, i.e. adding a line $\alpha = (\alpha_1, \alpha_2, \alpha_3)$ to A.

- *Let $M_3 = \{x_1 = 0\}$. We consider the matrix B obtained by adding the line $\alpha = (1, 0, 0)$ to A.*
  - *If $\xi_1 = 0$ condition (2.57) is satisfied.*
  - *If $\xi_2 = 0$, $\xi_1, \xi_3 \neq 0$, then $\log e^{\alpha_2 - \alpha_1} = 0 - 1 < 0$ and (2.57) is not satisfied. Indeed one can check that $\dfrac{\tau_2}{\tau_1} \notin \mathcal{G}_B$.*
  - *If $\xi_3 = 0$, $\xi_1, \xi_2 \neq 0$, then $\log e^{\alpha_2 - \alpha_1} = 0 - 1 < 0$ and (2.57) is not satisfied. Indeed one can check that $\dfrac{\tau_2}{\tau_1} \notin \mathcal{G}_B$.*
- *Let $M_3 = \{x_2 = 0\}$. We consider the matrix B obtained by adding the line $\alpha = (0, 1, 0)$ to A.*

- *If* $\xi_1 = 0,\ \xi_2, \xi_3 \neq 0$, *then* $\log e^{\alpha_1+\alpha_3-\alpha_2} = 0+0-1 < 0$ *and* (2.57) *is not satisfied. Indeed one can check that* $\frac{\tau_1\tau_3}{\tau_2} \notin \mathcal{G}_B$.
- *If* $\xi_2 = 0$ *condition* (2.57) *is satisfied.*
- *If* $\xi_3 = 0,\ \xi_1, \xi_2 \neq 0$, *then* $\log e^{\alpha_1+\alpha_3-\alpha_2} = 0+0-1 < 0$ *and* (2.57) *is not satisfied. Indeed one can check that* $\frac{\tau_1\tau_3}{\tau_2} \notin \mathcal{G}_B$.

- *Let* $M_3 = \{x_3 = 0\}$. *We consider the matrix* B *obtained by adding the line* $\alpha = (0,0,1)$ *to* A.
  - *If* $\xi_1 = 0,\ \xi_2, \xi_3 \neq 0$, *then* $\log e^{\alpha_2-\alpha_3} = 0-1 < 0$ *and* (2.57) *is not satisfied. Indeed one can check that* $\frac{\tau_2}{\tau_3} \notin \mathcal{G}_B$.
  - *If* $\xi_2 = 0,\ \xi_1, \xi_3 \neq 0$, *then* $\log e^{\alpha_2-\alpha_3} = 0-1 < 0$ *and* (2.57) *is not satisfied. Indeed one can check that* $\frac{\tau_2}{\tau_3} \notin \mathcal{G}_B$.
  - *If* $\xi_3 = 0$ *condition* (2.57) *is satisfied.*

- *Let* $M_3 = \{x_1 = x_3 = 0\}$. *We consider the matrix* B *obtained by adding the line* $\alpha = (1,0,1)$ *to* A.
  - *If* $\xi_1 = 0,\ \xi_2, \xi_3 \neq 0$, *then* $\log e^{\alpha_2-\alpha_3} = 0-1 < 0$ *and* (2.57) *is not satisfied. Indeed one can check that* $\frac{\tau_2}{\tau_3} \notin \mathcal{G}_B$.
  - *If* $\xi_2 = 0,\ \xi_1, \xi_3 \neq 0$, *then* $\log e^{\alpha_2-\alpha_1-\alpha_3} = 0-1-1 < 0$ *and* (2.57) *is not satisfied. Indeed one can check that* $\frac{\tau_2}{\tau_1\tau_3} \notin \mathcal{G}_B$.
  - *If* $\xi_3 = 0,\ \xi_1, \xi_2 \neq 0$, *then* $\log e^{\alpha_2-\alpha_1} = 0-1 < 0$ *and* (2.57) *is not satisfied. Indeed one can check that* $\frac{\tau_2}{\tau_1} \notin \mathcal{G}_B$.
  - *If* $\xi_1 = \xi_3 = 0$ *condition* (2.57) *is satisfied.*

- *Let* $M_3 = \{x_1 = x_2 = x_3 = 0\}$. *We consider the matrix* B *obtained by adding the line* $\alpha = (1,1,1)$ *to* A.
  - *If* $\xi_1 = 0$ *condition* (2.57) *is satisfied.*
  - *If* $\xi_2 = 0,\ \xi_1, \xi_3 \neq 0$, *then* $\log e^{\alpha_2-\alpha_1-\alpha_3} = 1-1-1 < 0$ *and* (2.57) *is not satisfied. Indeed one can check that* $\frac{\tau_2}{\tau_1\tau_3} \notin \mathcal{G}_B$.
  - *If* $\xi_3 = 0$ *condition* (2.57) *is satisfied.*

Example 2.57. *In* $\mathbb{R}^4$, *let us consider the family* $\chi_A = \{M_1, M_2\}$, *with* $M_1 = \{x_1 = x_2 = x_4 = 0\}$, $M_2 = \{x_2 = x_3 = 0\}$. *The matrix* A *is*

$$\begin{pmatrix} 1 & 1 & 0 & 1 \\ 0 & 1 & 1 & 0 \end{pmatrix}$$

If $\xi_3 = 0,\ \xi_1, \xi_2, \xi_4 \neq 0$, *we have*

$$F_A^1 = \left\{(\tau_1, 0), \left(\frac{\tau_2}{\tau_1}, 0\right), \left(\frac{\tau_1\tau_3}{\tau_2}, 0\right), \left(\frac{\tau_4}{\tau_1}, |\xi_4|\right), \left(\frac{\tau_1}{\tau_4}, \frac{1}{|\xi_4|}\right)\right\}.$$

We study the restriction when we consider $\chi_B = \{M_1, M_2, M_3\}$ *obtained by adding a manifold* M_3 *to* χ_A, *i.e. adding a line* $\alpha = (\alpha_1, \alpha_2, \alpha_3, \alpha_4)$ *to* A.

- *Let* $M_3 = \{x_2 = 0\}$. *We consider the matrix* B *obtained by adding the line* $\alpha = (0,1,0,0)$ *to* A. *Then* $\log e^{\alpha_1+\alpha_3-\alpha_2} = 0 + 0 - 1 < 0$ *and* (2.57) *is not satisfied. Indeed one can check that* $\left(\dfrac{\tau_1\tau_3}{\tau_2}, 0\right) \notin \mathcal{G}_B$.
- *Let* $M_3 = \{x_2 = x_4 = 0\}$. *We consider the matrix* B *obtained by adding the line* $\alpha = (0,1,0,1)$ *to* A. *Then* $\log e^{\alpha_1-\alpha_4} = 0 - 1 < 0$ *and* (2.57) *is not satisfied. Indeed one can check that* $\left(\dfrac{\tau_1}{\tau_4}, \dfrac{1}{|\xi_4|}\right) \notin \mathcal{G}_B$.
- *Let* $M_3 = \{x_1 = x_2 = x_3 = 0\}$. *We consider the matrix* B *obtained by adding the line* $\alpha = (1,1,1,0)$ *to* A. *Then* $\log e^{\alpha_1-\alpha_4} = 1 - 0 > 0$ *and* (2.57) *is not satisfied. Indeed one can check that* $\left(\dfrac{\tau_1}{\tau_4}, 0\right) \in \mathcal{G}_B$.
- *Let* $M_3 = \{x_1 = x_2 = x_3 = x_4 = 0\}$. *We consider the matrix* B *obtained by adding the line* $\alpha = (1,1,1,1)$ *to* A. *Then* (2.57) *is satisfied.*

3 The functor of multi-specialization

In this section we extend the notion of multi-specialization of [9] to our new setting. References are made to [14] for the classical theory of specialization along a submanifold.

3.1 A decomposition result on the normal deformation

Definition 3.1. *Denote by* $\mathrm{Op}(\widetilde{X})$ *the category of open subsets of* $\widetilde{X}$, *and let* Z *be a subset of* $\widetilde{X}$.

(i) We set $\mathbb{R}^+_j Z = \mu_j(Z, \mathbb{R}^+)$. *If* $U \in \mathrm{Op}(\widetilde{X})$, *then* $\mathbb{R}^+_j U \in \mathrm{Op}(\widetilde{X})$ *since* μ_j *is open for each* $j = 1, \dots, \ell$.

(ii) Let $J = \{j_1, \dots, j_k\} \subset \{1, \dots, \ell\}$. *We set*

$$\mathbb{R}^+_J Z = \mathbb{R}^+_{j_1} \cdots \mathbb{R}^+_{j_k} Z = \mu_{j_1}(\cdots \mu_{j_k}(Z, \mathbb{R}^+), \dots, \mathbb{R}^+).$$

We set $(\mathbb{R}^+)^\ell Z = \mathbb{R}^+_{\{1,\dots,\ell\}} Z = \mu(Z, (\mathbb{R}^+)^\ell)$. *If* $U \in \mathrm{Op}(\widetilde{X})$, *then* $\mathbb{R}^+_J U \in \mathrm{Op}(\widetilde{X})$ *since* μ_j *is open for each* $j \in \{1, \dots, \ell\}$.

(iii) We say that Z *is* $(\mathbb{R}^+)^\ell$*-conic (*ℓ*-conic for short) if* $Z = (\mathbb{R}^+)^\ell Z$. *In other words,* Z *is invariant by the action of* μ_j, $j = 1, \dots, \ell$.

Definition 3.2. *(i) We say that a subset* Z *of* $\widetilde{X}$ *is* $\mathbb{R}^+_j$*-connected if* $Z \cap \mathbb{R}^+_j x$ *is connected for each* $x \in Z$.

(ii) We say that a subset Z *of* $\widetilde{X}$ *is* $(\mathbb{R}^+)^\ell$*-connected if there exists a permutation*

$\sigma : \{1, \dots, \ell\} \to \{1, \dots, \ell\}$ *such that*

$$\begin{cases} Z \text{ is } \mathbb{R}^+_{\sigma(1)}\text{-connected}, \\ \mathbb{R}^+_{\sigma(1)} Z \text{ is } \mathbb{R}^+_{\sigma(2)}\text{-connected}, \\ \vdots \\ \mathbb{R}^+_{\sigma(1)} \cdots \mathbb{R}^+_{\sigma(\ell-1)} Z \text{ is } \mathbb{R}^+_{\sigma(\ell)}\text{-connected}. \end{cases}$$

The proof of the following is almost the same as that of Proposition 4.1.3 of [14], and we shall not repeat it. For an open subset $T \subset \Omega$, we set

$$O_{\overline{\Omega} \setminus S}(T) = (\overline{\Omega} \setminus S) \setminus \overline{(\Omega \setminus T)} \subset \overline{\Omega} \setminus S,$$

which is nothing but the maximal open subset W in $\overline{\Omega} \setminus S$ such that $W \cap \Omega = T$.

Proposition 3.3. *Let V be an $(\mathbb{R}^+)^\ell$-conic open subset of the zero section S_χ.*

(i) Let W be an open neighborhood of V in $\widetilde{X}$, and let $U = \widetilde{p}(W \cap \Omega)$. Then $V \cap C_\chi(X \setminus U) = \emptyset$.

(ii) Let U be an open subset of X such that $V \cap C_\chi(X \setminus U) = \emptyset$. Then $W = O_{\overline{\Omega} \setminus S}(\widetilde{p}^{-1}(U)) \cup V$ is an open neighborhood of V in $\overline{\Omega}$. Note that W satisfies $W \cap S = V$ and $W \cap \Omega = \widetilde{p}^{-1}(U)$.

Set, for $j = 1, \dots, \ell$,

$$S_{\chi,j} := \{(x^{(0)}, \xi^{(1)}, \dots, \xi^{(\ell)}) \in S_\chi;\ \xi^{(j)} = 0\} \subset S_\chi$$

and $\pi_j : S_\chi \to S_{\chi,j}$ be the canonical projection. Let V be an $(\mathbb{R}^+)^\ell$-conic subanalytic subset of S_χ. We introduce the following conditions Va. and Vb. of V for each j.

Va. V does not intersect $S_{\chi,j}$.

Vb. $\pi_j(V) \subset \pi_j(V \cap S_{\chi,j})$.

Then we have the following proposition whose proof will be given in Appendix A.3.

Proposition 3.4. *Let V be an $(\mathbb{R}^+)^\ell$-conic subanalytic subset of S_χ which satisfies the condition either Va. or Vb. for each j. Assume that $X = \mathbb{R}^n$ with coordinates $(x^{(0)}, x^{(1)}, \dots, x^{(m)})$ defined by χ. Then any subanalytic neighborhood W of V in $\widetilde{X}$ contains $\widetilde{W}$ open and subanalytic in $\widetilde{X}$ such that:*

$$\begin{cases} (i)\ \widetilde{W} \cap \Omega \text{ is } (\mathbb{R}^+)^\ell\text{-connected}, \\ (ii)\ \mathbb{R}^+_1 \cdots \mathbb{R}^+_\ell (\widetilde{W} \cap \Omega) = \widetilde{p}^{-1}(\widetilde{p}(W \cap \Omega)) \text{ is subanalytic in } \widetilde{X}. \end{cases} \tag{3.1}$$

Let $Z \subset S_\chi$ be closed conic and subanalytic. We have the following corollary.

Corollary 3.5. *Let V be an $(\mathbb{R}^+)^\ell$-conic open subanalytic subset of Z. Then there exists a locally finite family $\{V^{(\alpha)}\}_\alpha$ of $(\mathbb{R}^+)^\ell$-conic open subanalytic subsets in Z which satisfies the following conditions.*

1. *$V = \bigcup_\alpha V^{(\alpha)}$ and each $V^{(\alpha)}$ satisfies the condition either Va. or Vb. for each j.*
2. *For any subanalytic open neighborhood W of V in $\widetilde{X}$, there exists a subanalytic open neighborhood $\widetilde{W}^{(\alpha)} \subset W$ of $V^{(\alpha)}$ for which (3.1) of Proposition 3.4 holds. Furthermore $\{\widetilde{p}(\widetilde{W}^{(\alpha)} \cap \Omega)\}_\alpha$ is a locally finite family of subanalytic open subsets of X.*

3.2 Multi-specialization

Let k be a field and denote by $\mathrm{Mod}(k_{X_{sa}})$ (resp. $D^b(k_{X_{sa}})$) the category (resp. bounded derived category) of sheaves on the subanalytic site X_{sa}. For the theory of sheaves on subanalytic sites we refer to [16] and [20]. For classical sheaf theory we refer to [14]. For an exposition on $(\mathbb{R}^+)^\ell$-conic sheaves see the appendix.

Definition 3.6. *The multi-specialization along χ is the functor*

$$\nu_\chi^{sa} : D^b(k_{X_{sa}}) \to D^b(k_{S_{\chi sa}}), \qquad F \mapsto s^{-1} R\Gamma_\Omega p^{-1} F.$$

It follows from the definition (see also the proof of Theorem 6.3 (i) of [9]) that $\nu_\chi^{sa} F$ is conic, i.e., it belongs to $D^b_{(\mathbb{R}^+)^\ell}(k_{S_{\chi sa}})$. The following result can be obtained adapting the proof of Lemma 6.1 of [9].

Lemma 3.7. *Let $F \in D^b(k_{X_{sa}})$. There is a natural isomorphism*

$$s^{-1} R\Gamma_\Omega p^{-1} F \simeq s^! (p^{-1} F)_\Omega.$$

Thanks to Proposition 3.4 we can give a description of the sections of the multi-specialization of $F \in D^b(k_{X_{sa}})$.

Theorem 3.8. *Let V be a conic subanalytic open subset of S_χ satisfying either condition Va. or Vb. for each j. Then:*

$$H^j(V; \nu_\chi^{sa} F) \simeq \varinjlim_U H^j(U; F),$$

where U ranges through the family of open subanalytic subsets of X such that $C_\chi(X \setminus U) \cap V = \emptyset$.

Proof. Let $U \in \mathrm{Op}(X_{sa})$ such that $V \cap C_\chi(X \setminus U) = \emptyset$. We have the chain of morphisms

$$\begin{aligned} R\Gamma(U; F) &\to R\Gamma(p^{-1}(U); p^{-1}F) \\ &\to R\Gamma(p^{-1}(U) \cap \Omega; p^{-1}F) \\ &\to R\Gamma(O_{\overline{\Omega} \setminus S}(\widetilde{p}^{-1}(U)) \cup V; R\Gamma_\Omega p^{-1}F) \\ &\to R\Gamma(V; \nu_\chi^{sa} F), \end{aligned}$$

where the third arrow exists since $O_{\overline{\Omega}\setminus S}(\widetilde{p}^{-1}(U)) \cup V$ is a neighborhood of V in $\overline{\Omega}$ by Proposition 3.3 (ii).

Let us show that it is an isomorphism. We have

$$\begin{aligned} H^k(V;\nu^{sa}_\chi F) &\simeq \varinjlim_W H^k(W; R\Gamma_\Omega p^{-1}F) \\ &\simeq \varinjlim_W H^k(W\cap\Omega; p^{-1}F), \end{aligned}$$

where W ranges through the family of subanalytic open neighborhoods of V in $\widetilde{X}$. By Proposition 3.4, we may assume that W satisfies (3.1). Since $p^{-1}F$ is $(\mathbb{R}^+)^\ell$-conic, we have

$$\begin{aligned} H^k(W\cap\Omega; p^{-1}F) &\simeq H^k(p^{-1}(p(W\cap\Omega)); p^{-1}F) \\ &\simeq H^k(p(W\cap\Omega)\times\{(1)_\ell\}; p^{-1}F) \\ &\simeq H^k(p(W\cap\Omega); F), \end{aligned}$$

where $(1)_\ell = (1,\dots,1)\in\mathbb{R}^\ell$. The second isomorphism follows since every subanalytic neighborhood of $p(W\cap\Omega)\times\{(1)_\ell\}$ contains an $(\mathbb{R}^+)^\ell$-connected subanalytic neighborhood (the proof is similar to that of Proposition 3.4). By Proposition 3.3 (i) we have that $p(W\cap\Omega)$ ranges through the family of subanalytic open subsets U of X such that $V\cap C_\chi(X\setminus U)=\emptyset$ and we obtain the result. □

Remember that a sheaf $F\in \mathrm{Mod}(k_{X_{sa}})$ is said to be quasi-injective if the restriction morphism $\Gamma(U;F)\to\Gamma(V;F)$ is surjective for each $U,V\in \mathrm{Op}^c(X_{sa})$ with $U\supseteq V$. Since quasi-injective sheaves are $\Gamma(U;\cdot)$-acyclic for each $U\in\mathrm{Op}(X_{sa})$ we get the following result.

Corollary 3.9. *Let $F\in\mathrm{Mod}(k_{X_{sa}})$ be quasi-injective. Then F is ν^{sa}_χ-acyclic.*

Proof. It is enough to prove that $H^k(V;\nu^{sa}_\chi F) = 0$, $k\neq 0$, on a basis for the topology of $S_{\chi sa}$. Being $\nu^{sa}_\chi F$ conic, we may assume that V is conic as well. By Corollary 3.5 we may assume that V satisfies the condition either Va. or Vb. for each j. Thanks to Theorem 3.8 it is enough to show that $H^k(U;F)=0$, $k\neq 0$ when $U\in\mathrm{Op}(X_{sa})$. Since quasi-injective sheaves are $\Gamma(U;\cdot)$-acyclic for each $U\in\mathrm{Op}(X_{sa})$ we get the result. □

Let $p=(x^{(0)};\xi^{(1)},\dots,\xi^{(\ell)})\in S_\chi$, let $\pi: S_\chi\to\bigcap_{j=1}^\ell M_j$ be the projection. Namely, $\pi(x^{(0)};\xi^{(1)},\dots,\xi^{(\ell)}) = (x^{(0)},0^{(1)},\dots,0^{(\ell)})$. Let B_ϵ be an open ball of radius $\epsilon>0$ with its center at $\pi(p)$ and set

$$\mathrm{Cone}_\chi(p,\epsilon) := \{V\cap B_\epsilon;\ V\in C(\xi,F^q)\}.$$

Applying the functor $\rho^{-1}\colon D^b(k_{S_{\chi sa}}) \to D^b(k_{S_\chi})$ (see [20] for details) and Theorem 3.8 (see also Corollary 6.5 of [9]) we can calculate the fibers at $p \in S_\chi$ which are given by

$$(H^j \rho^{-1} \nu_\chi^{sa} F)_p \simeq \varinjlim_{W} H^j(W; F), \tag{3.2}$$

where W ranges through the family $\mathrm{Cone}_\chi(p, \epsilon)$ for $\epsilon > 0$.

From Proposition 2.52 we can deduce a condition for the restriction of multi-specialization. Set $\chi_A = \{M_1, \dots, M_\ell\}$, $\chi_B = \chi_A \cup \{M_{\ell+1}\}$. We follow the notations of Section 2. We assume that the first ℓ (resp. $\ell + 1$) columns of the matrix A (resp. B) are linearly independent. We omit the coordinate $x^{(0)}$ to lighten notations. Given $\xi = (\xi^{(1)}, \dots, \xi^{(m)}) \in \mathbb{R}^n$ we will call $p_A \in S_A$, the point $p_A = (\xi^{(1)}, \dots, \xi^{(m)}; 0 \dots, 0) \in \mathbb{R}^{n+\ell}$, and $p_B \in S_B$, the point $p_B = (\xi^{(1)}, \dots, \xi^{(m)}; 0 \dots, 0) \in \mathbb{R}^{n+\ell+1}$.

Corollary 3.10. *Let $F \in D^b(k_{X_{sa}})$. Let $K \subseteq \{1, \dots, m\}$. Suppose that on $S_K = \{\xi^{(k)} = 0,\ k \in K\}$ conditions (2.58) are satisfied. Then we have*

$$(\nu^{sa}_{\chi_A} F)|_{S_K} \simeq (\nu^{sa}_{\chi_B} F)|_{S_K}.$$

3.3 Operations

Let X and Y be real analytic manifolds and let $\chi^M = \{M_j\}_{j=1}^{\ell}$, $\chi^N = \{N_j\}_{j=1}^{\ell}$ be families of smooth closed submanifolds of X and Y respectively. Let $f : X \to Y$ be a morphism of real analytic manifolds such that $f(M_j) \subseteq N_j$, $j = 1, \dots, \ell$.

Remark 3.11. *In order to study more general situations, we will assume $X = Y = \mathbb{R}^n$ and consider the local model. Even if we work locally, the theory below completely covers the manifold case.*

Remark 3.12. *Let σ_A be a positive rational number so that $a_{jk} \in \mathbb{Z}$ and all the a_{jk} have no common divisors. Let $\mu_{\sigma_A A_\chi}$ be the action on X associated with the matrix $\sigma_A A_\chi$. Since multi-cones associated with the action μ and those with the action $\mu_{\sigma_A A_\chi}$ determine equivalent families, we may assume from the beginning that all the entries of A_χ are non-negative integers and $\sigma_A = 1$.*

We want to extend f to a morphism $\widetilde{f} : \widetilde{X} \to \widetilde{Y}$. This is done by repeatedly employing the usual construction of a morphism between normal deformations, i.e. we extend f to $\widetilde{f}_1 : \widetilde{X}_{M_1} \to \widetilde{Y}_{N_1}$, then we extend $\widetilde{f}_1$ to $\widetilde{f}_{1,2} : \widetilde{X}_{M_1,M_2} \to \widetilde{Y}_{N_1,N_2}$. Then we can define recursively $\widetilde{f} : \widetilde{X} \to \widetilde{Y}$ by extending the morphism $\widetilde{f}_{1,\dots,\ell-1} : \widetilde{X}_{M_1,\dots,M_{\ell-1}} \to \widetilde{Y}_{N_1,\dots,N_{\ell-1}}$ to the normal deformations with respect to M_ℓ and N_ℓ respectively. We also denote by S^M (resp. S^N) the zero section of $\widetilde{X}$ (resp. $\widetilde{Y}$). In this general case we have to put some conditions to assure that $\widetilde{f}$ is well defined on the zero section.

In a local coordinate system set

$$\begin{aligned} x &= (x^{(0)}, x^{(1)}, \dots, x^{(m_X)}), \\ y &= (y^{(0)}, y^{(1)}, \dots, y^{(m_Y)}), \\ f(x) &= (f^{(0)}(x), f^{(1)}(x), \dots, f^{(m_Y)}(x)). \end{aligned}$$

Hereafter we omit the coordinates blocks $x^{(0)}$ and $y^{(0)}$ for simplicity. Outside $\{t_1 t_2 \dots t_\ell = 0\}$ in $\widetilde{X}$, the morphism $\widetilde{f} : \widetilde{X} \to \widetilde{Y}$

$$(x_1, \dots, x_n, t_1, \dots, t_\ell) \to (y_1, \dots, y_m, \hat{t}_1, \dots, \hat{t}_\ell)$$

is explicitly written in the form

$$\begin{cases} \varphi_k^N(\hat{t}) y^{(k)} = f^{(k)}(\varphi_1^M(t) x^{(1)}, \dots, \varphi_{m_X}^M(t) x^{(m_X)}) & (k = 1, \dots, m_Y), \\ \hat{t}_j = t_j \qquad (j = 1, \dots, \ell), \end{cases}$$

which naturally extends to the whole $\widetilde{X}$ if the limit

$$\lim_{t \to 0} \frac{f^{(k)}(\varphi_1^M(t) x^{(1)}, \dots, \varphi_{m_X}^M(t) x^{(m_X)})}{\varphi_k^N(t)}$$

exists for each $k = 1, \dots, m_Y$. In order to understand the restriction of $\widetilde{f}$ to the whole $\widetilde{X}$, we set

$$I = \bigcup_{j \in J_k^N} I_j^M, \ M_I = \bigcap_{j \in J_k^N} M_j,$$

where J_k^N denotes the set $\{j \in \{1, \dots, \ell\}, \hat{I}_k^N \subset I_j^N\}$. By expanding $f^{(k)}(x)$ along the submanifold M_I, we obtain

$$f^{(k)}(x) = \sum_{i \in I} \left. \frac{\partial f^{(k)}}{\partial x^{(i)}} \right|_{M_I} x^{(i)} + \frac{1}{2} \sum_{i_1, i_2 \in I} \left. \frac{\partial^2 f^{(k)}}{\partial x^{(i_1)} \partial x^{(i_2)}} \right|_{M_I} x^{(i_1)} x^{(i_2)} + \dots,$$

as $f_k|_{M_I} = 0$ holds. Then we get, on $t_1 \dots t_\ell \neq 0$,

$$y_k = \sum_{i \in I} \left. \frac{\partial f^{(k)}}{\partial x^{(i)}} \right|_{M_I} \frac{\varphi_i^M(t)}{\varphi_k^N(t)} x^{(i)} + \frac{1}{2} \sum_{i_1, i_2 \in I} \left. \frac{\partial^2 f^{(k)}}{\partial x^{(i_1)} \partial x^{(i_2)}} \right|_{M_I} \frac{\varphi_{i_1}^M(t) \varphi_{i_2}^M(t)}{\varphi_k^N(t)} x^{(i_1)} x^{(i_2)} + \dots.$$

Note that the limit for $t \to 0$ exists under the condition

$$\left. \frac{\partial f^{(k)}}{\partial x^{(i_1)} \dots \partial x^{(i_p)}} \right|_{M_I} = 0 \quad \text{if } \varphi_k^N \nmid \varphi_{i_1}^M \cdots \varphi_{i_p}^M. \tag{3.3}$$

Remark 3.13. *If there exists $j \in J_k^N$ such that $\{i_1, \dots, i_p\} \cap I_j^M = \emptyset$, as $f^{(k)}|_{M_j} = 0$, we have*

$$\left. \frac{\partial f^{(k)}}{\partial x^{(i_1)} \dots \partial x^{(i_p)}} \right|_{M_I} = 0.$$

If the matrix A is associated with χ, i.e., its entries are either 0 or 1,

$$\varphi_k^N \nmid \varphi_{i_1}^M \cdots \varphi_{i_p}^M \Rightarrow \exists\, j \in J_k^N \text{ such that } \{i_1, \dots, i_p\} \cap I_j^M = \emptyset.$$

Then (3.3) always holds.

Suppose that $\varphi_k^N \mid \varphi_{i_1}^M \cdots \varphi_{i_p}^M$.

(i) If there exists $j \in J_k^N$ such that $\{i_1, \dots, i_p\} \cap I_j^M = \emptyset$, we have

$$\left.\frac{\partial f^{(k)}}{\partial x^{(i_1)} \cdots \partial x^{(i_p)}}\right|_{M_I} = 0.$$

(ii) If $\alpha_{jk}^N < \alpha_{ji_1}^M + \cdots + \alpha_{ji_p}^M$, for some $j \in \{1, \dots, \ell\}$, we have

$$\lim_{t \to 0} \frac{\varphi_{i_1}^M(t) \cdots \varphi_{i_p}^M(t)}{\varphi_k^N(t)} \to 0.$$

(iii) If $\alpha_{jk}^N = \alpha_{ji_1}^M + \cdots + \alpha_{ji_p}^M$, for all $j \in \{1, \dots, \ell\}$, we have

$$\lim_{t \to 0} \frac{\varphi_{i_1}^M(t) \cdots \varphi_{i_p}^M(t)}{\varphi_k^N(t)} \to 1.$$

By the above observations, assuming (3.3), in order to compute the restriction to the zero section we are interested in sequences $P = \{i_1, \dots, i_{\sharp P}\}$ ($i_p \in \{1, \dots, m_X\}$ $(p = 1, \dots, \sharp P)$) of indices of finite length, such that

$$\alpha_{jk}^N = \sum_{p=1}^{\sharp P} \alpha_{ji_p}^M \qquad (j = 1, \dots, \ell). \tag{3.4}$$

Let us consider the family $\mathcal{F}_k$ of sequences P of indices of finite length satisfying (3.4). The morphism $\widetilde{f}$ is described by, on $S \subset \widetilde{X}$,

$$y_k = \sum_{P \in \mathcal{F}_k} \frac{1}{\sharp P!} \left.\frac{\partial^{\sharp P} f^{(k)}}{\partial x^{(i_1)} \cdots \partial x^{(i_{\sharp P})}}\right|_M x^{(i_1)} \cdots x^{(i_{\sharp P})} \qquad (k = 1, \dots, m_Y).$$

Here $M := M_1 \cap \cdots \cap M_\ell$.

Remark 3.14. *The order is important in $\mathcal{F}_k$ as it is the set of sequences. For example, $\{1, 2\} \neq \{2, 1\}$.*

Remark 3.15. *When $m_X = \ell$, i.e. χ^M is of normal type, the family $\mathcal{F}_k$ consists of at most one element. This follows from the invertibility of the matrix associated with χ^M.*

Example 3.16. *Let $X = Y = \mathbb{R}^3$, $M_1 = \{x_1 = x_2 = 0\}$, $M_2 = \{x_2 = x_3 = 0\}$, $N_1 = 0$, $N_2 = \{x_2 = x_3 = 0\}$, $f(x_1, x_2, x_3) = (x_1, x_1x_3 + x_2, x_1x_3)$. Then*

$$\varphi_1^M = t_1, \quad \varphi_2^M = t_1t_2, \quad \varphi_3^M = t_2,$$
$$\varphi_1^N = t_1, \quad \varphi_2^N = t_1t_2, \quad \varphi_3^N = t_1t_2.$$

Condition (3.3) *is satisfied and we have*

$$\begin{aligned} y_1 &= \left.\frac{\partial f_1}{\partial x_1}\right|_0 x_1 = x_1, \\ y_2 &= \left.\frac{\partial f_2}{\partial x_2}\right|_0 x_2 + \frac{1}{2}\left.\frac{\partial^2 f_2}{\partial x_1 \partial x_3}\right|_0 x_1x_3 + \frac{1}{2}\left.\frac{\partial^2 f_2}{\partial x_3 \partial x_1}\right|_0 x_3x_1 = x_2 + x_1x_3, \\ y_3 &= \frac{1}{2}\left.\frac{\partial^2 f_3}{\partial x_1 \partial x_3}\right|_0 x_1x_3 + \frac{1}{2}\left.\frac{\partial^2 f_2}{\partial x_3 \partial x_1}\right|_0 x_3x_1 = x_1x_3. \end{aligned}$$

Note that, as stressed in Remark 3.14, the order of the elements P in $\mathcal{F}_2$ and $\mathcal{F}_3$ is important.

Example 3.17. *Let $X = Y = \mathbb{R}^2$, $M_1 = \{x_1 = 0\}$, $M_2 = \{x_2 = 0\}$, $N_1 = N_2 = \{0\}$,*

$$A_{\chi^M} = \begin{pmatrix} 1 & 0 \\ 0 & 1 \end{pmatrix}, \qquad A_{\chi^N} = \begin{pmatrix} 3 & 2 \\ 1 & 1 \end{pmatrix},$$

$f(x_1, x_2) = (x_1^3x_2, x_1^2x_2)$. *Then*

$$\begin{aligned} \varphi_1^M &= t_1, \quad \varphi_2^M = t_2, \\ \varphi_1^N &= t_1^3t_2, \quad \varphi_2^N = t_1^2t_2. \end{aligned}$$

Condition (3.3) *is satisfied and we have*

$$\begin{aligned} y_1 &= \frac{1}{4!}\cdot 4 \cdot \left.\frac{\partial^4 f_1}{\partial x_1^3 \partial x_2}\right|_0 x_1^3x_2 = x_1^3x_2, \\ y_2 &= \frac{1}{3!}\cdot 3 \cdot \left.\frac{\partial^3 f_1}{\partial x_1^2 \partial x_2}\right|_0 x_1^2x_2 = x_1^2x_2. \end{aligned}$$

Note that we multiplied the first (resp. second) line by 4 (resp. 3) taking in account the order of the elements P in $\mathcal{F}_1$ (resp. $\mathcal{F}_2$).

Suppose that f satisfies (3.3) and $f(M_i) \subseteq N_i$, $i = 1, \dots, \ell$. We call $T_\chi f$ the induced map on the zero section. Let $J = \{j_1, \dots, j_k\}$ be a non-empty subset in $\{1, \dots, \ell\}$ and set $\chi_J^M = \{M_{j_1}, \dots, M_{j_k}\}$. We also set

$$\mathbb{R}_{>0}^{J^c} = \{t \in \mathbb{R}^\ell;\, t_j = 0\,(j \in J),\, t_j > 0\,(j \notin J)\} \subset \mathbb{R}^\ell.$$

The map $\widetilde{T}_{\chi_J} f$ denotes the restriction of $\widetilde{f}$ to $\{t \in \mathbb{R}_{>0}^{J^c}\}$, that is, it is a map from $\widetilde{X} \times_{\mathbb{R}^\ell} \mathbb{R}_{>0}^{J^c}$ to $\widetilde{Y} \times_{\mathbb{R}^\ell} \mathbb{R}_{>0}^{J^c}$. In the following we will denote with the same symbol $C_{\chi_J^M}(Z)$ the normal cone with respect to χ_J^M and its inverse image via the map $\widetilde{p}_{\chi_{J^c}^M} : \widetilde{X} \times_{\mathbb{R}^\ell} \mathbb{R}_{>0}^{J^c} \to S_{\chi_J^M}$ (see also the remark below for the map $\widetilde{p}_{\chi_{J^c}^M}$).

Proposition 3.18. *Let $F \in D^b(k_{X_{sa}})$.*

(i) There exists a commutative diagram of canonical morphisms

$$\begin{array}{ccc} R(T_\chi f)_{!!}\nu^{sa}_{\chi^M}F & \longrightarrow & \nu^{sa}_{\chi^N}Rf_{!!}F \\ \downarrow & & \downarrow \\ R(T_\chi f)_*\nu^{sa}_{\chi^M}F & \longleftarrow & \nu^{sa}_{\chi^N}Rf_*F. \end{array}$$

(ii) Moreover if f is proper over supp$F$ *and $\widetilde{T}_{\chi_J}f$ is proper over $C_{\chi_J^M}(\mathrm{supp}F)$ for each (non-empty) $J \subset \{1,\dots,\ell\}$, and if* $\mathrm{supp}F \cap f^{-1}(N_j) \subseteq M_j$, $j \in \{1,\dots,\ell\}$, *then the above morphisms are isomorphisms.*

Note that the condition in (ii) of the above proposition that $\widetilde{T}_{\chi_J}f$ is proper over $C_{\chi_J^M}(\mathrm{supp}F)$ is equivalently saying that $T_{\chi_J}f : S_{\chi_J^M} \to S_{\chi_J^N}$ is proper over $C_{\chi_J^M}(\mathrm{supp}F)$.

In fact, set $d = \ell - \#J$, $t' = (t_j)_{j\notin J}$ and $t'' = (t_{j_k})_{j_k\in J}$. Then $\widetilde{p}_{\chi^M_{J^c}}(x;t')$ is given by $p_\chi(x;t',1'')$, where $p_\chi : \widetilde{X} \to X$ is the multi-normal deformation along χ and $1'' = (1,\cdots,1)$. Furthermore, we have the following commutative diagram whose vertical arrows are all isomorphisms:

$$\begin{array}{ccc} \widetilde{X} \times_{\mathbb{R}^\ell} \mathbb{R}^{J^c}_{>0} & \xrightarrow{\widetilde{T}_{\chi_J}f} & \widetilde{Y} \times_{\mathbb{R}^\ell} \mathbb{R}^{J^c}_{>0} \\ \downarrow & & \downarrow \\ S_{\chi_J^M} \times (\mathbb{R}^+)^d & \xrightarrow{(T_{\chi_J}f)\times \mathrm{id}_{(\mathbb{R}^+)^d}} & S_{\chi_J^N} \times (\mathbb{R}^+)^d. \end{array}$$

Here the left and right vertical arrows are given by $(x;t') \mapsto (\widetilde{p}_{\chi^M_{J^c}}(x;t');t')$ and $(y;t') \mapsto (\widetilde{p}_{\chi^N_{J^c}}(x;t');t')$, respectively.

Let $Z \subset S_{\chi_J^M}$ be a closed subset. For any compact set $K \subset \widetilde{Y} \times_{\mathbb{R}^\ell} \mathbb{R}^{J^c}_{>0}$, we can find $\delta > 0$ such that $K \subset \{t' \in [\delta,\delta^{-1}]^d,\ t'' = 0\}$. Further $(\widetilde{p}_{\chi^M_{J^c}})^{-1}(Z)$ is identified with $Z \times (\mathbb{R}^+)^d$ in $S_{\chi_J^M} \times (\mathbb{R}^+)^d$ via the left vertical arrow.

Hence it follows from the above commutative diagram that the set

$$(\widetilde{p}_{\chi^M_{J^c}})^{-1}(Z) \cap (\widetilde{T}_{\chi_J}f)^{-1}(K) \subset \widetilde{X}$$

becomes compact for any compact set $K \subset \widetilde{Y} \times_{\mathbb{R}^\ell} \mathbb{R}^{J^c}_{>0}$ if and only if $Z \cap (T_{\chi_J}f)^{-1}(L) \subset S_{\chi_J^M}$ is compact for any compact set $L \subset S_{\chi_J^N}$. This shows the equivalence of the above two conditions.

Now we will give the proof of the proposition:

Proof. (i) The existence of the arrows is done as in [14] Proposition 4.2.4.

(ii) If $\widetilde{f}$ is proper over $\overline{\widetilde{p}^{-1}(\mathrm{supp}F)}$, then all the morphisms are isomorphisms. We have to prove that for a closed subset Z of X, the restriction of $\widetilde{f}$ to $\overline{\widetilde{p}^{-1}(Z)}$ is proper, if $Z \to Y$ and $\widetilde{T}_{\chi_J} f : C_{\chi_J^M}(Z) \to \widetilde{Y} \times_{\mathbb{R}^\ell} \mathbb{R}^{J^c}_{>0}$ for any non-empty J are proper, and if $Z \cap f^{-1}(N_j) \subseteq M_j$, $j \in \{1, \dots, \ell\}$. We argue by induction on $\sharp\chi^M$. If $\sharp\chi^M = 0$ the proof is trivial. Suppose it is true for $\sharp\chi^M \leq \ell - 1$. It follows from the hypothesis that the fibers of $\widetilde{f}$ restricted to $\overline{\widetilde{p}^{-1}(Z)}$ are compact (if $t_{j_k} = 0$, $j_k \in J$, and $t_j \neq 0$, $j \notin J$, this is a consequence of the fact that $\widetilde{T}_{\chi_J} f : C_{\chi_J^M}(Z) \to \widetilde{Y} \times_{\mathbb{R}^\ell} \mathbb{R}^{J^c}_{>0}$ is proper). Then it remains to prove that it is a closed map. Let $\{u_n\}_{n \in \mathbb{N}}$ be a sequence in $\overline{\widetilde{p}^{-1}(Z)}$ and suppose that $\{\widetilde{f}(u_n)\}_{n \in \mathbb{N}}$ converges. We shall find a convergent subsequence of $\{u_n\}_{n \in \mathbb{N}}$. We may also assume that $\{\widetilde{p}(u_n)\}_{n \in \mathbb{N}}$ converges as f is proper over Z. The map $\widetilde{p}^{-1}(Z) \setminus \{t_1 = \cdots = t_\ell = 0\} \to \widetilde{Y} \setminus \{t_1 = \cdots = t_\ell = 0\}$ is proper. Indeed, let K be a compact subset of $\widetilde{Y} \setminus \{t_1 = \cdots = t_\ell = 0\}$, and reduce to the case that K is contained in $\{c \leq t_j \leq d\} \subset \widetilde{Y}$, $c, d > 0$, for some $j \in \{1, \dots, \ell\}$. Suppose without loss of generality that $j = 1$. Then $K_1 := \widetilde{p}_{N_1}(K)$ is a compact subset of $\widetilde{Y}_{N_2 \dots N_\ell}$. Let us identify K_1 with $\widetilde{p}_{N_1}^{-1}(K_1) \cap \{t_1 = 1\}$. Then $\widetilde{f}^{-1}(K_1)$ is compact by the induction hypothesis. Hence $\widetilde{f}^{-1}(K) \subseteq \mu_1(\widetilde{f}^{-1}(K_1), [c, d])$ is compact since it is closed and contained in a compact subset. We may assume that $\{\widetilde{f}(u_n)\}_{n \in \mathbb{N}}$ converges to a point of $\{t_1 = \cdots = t_\ell = 0\}$. Then $\{\widetilde{p}(u_n)\}_{n \in \mathbb{N}}$ converges to a point of $Z \cap f^{-1}(N_1 \cap \cdots \cap N_\ell) \subseteq M_1 \cap \cdots \cap M_\ell$.

Taking local coordinates systems of X and Y, let

$$u_n = (x_n^{(1)}, \dots, x_n^{(m)}; t_{1n}, \dots, t_{\ell n}), \quad t_{jn} > 0, \ j = 1, \dots, \ell.$$

Then $t_{jn} \to 0$, $j = 1, \dots, \ell$ and $\varphi_i^M(t_n) x_n^{(i)} \to 0$, $i = 1, \dots, m$. It is enough to show that $\{|x_n^{(i)}|\}_{n \in \mathbb{N}}$ is bounded for each $i = 1, \dots, m$. We argue by contradiction. Up to make a permutation of $\{1, \dots, m\}$, we may suppose without loss of generality that $|x_n^{(i)}| \to +\infty$ for $0 < i \leq m_0 \leq m$.

Suppose we can find a non infinitesimal bounded sequence $\{\widetilde{u}_n\}_{n \in \mathbb{N}} \in \widetilde{p}^{-1}(Z)$ such that $\{\widetilde{f}(\widetilde{u}_n)\}_{n \in \mathbb{N}}$ converges to 0. Up to extract a subsequence we may assume that $\{u_n\}_{n \in \mathbb{N}}$ converges to a non zero vector v. Then $\widetilde{f}(w) = 0$ for each $w \in (\mathbb{R}^+)^\ell v$ which contradicts the fact that the fibers of $\overline{\widetilde{p}^{-1}(Z)} \to \widetilde{Y}$ are compact.

Suppose that there exist sequences $\{c_{jn}\}_{n \in \mathbb{N}}$, $j = 1, \dots, \ell$, satisfying the following properties:

(i) $c_{jn} \to +\infty$, $j = 1, \dots, \ell$,

(ii) $\displaystyle\prod_{j=1}^{\ell} c_{jn}^{\alpha_{m_0 j}} = |x_n^{(m_0)}|$,

(iii) $t_{jn} c_{jn} \to 0$, $j = 1, \dots, \ell$.

Let us construct the sequence $\{\overline{u}_n\}_{n\in\mathbb{N}}$, $\overline{u}_n = (\overline{x}_n^{(1)}, \dots, \overline{x}_n^{(m)}, \overline{t}_{1n}, \dots, \overline{t}_{\ell n})$ as follows:

$$\overline{x}_n^{(i)} = \frac{x_n^{(i)}}{\prod_{j=1}^{\ell} c_{jn}^{\alpha_{ij}}}, \qquad \overline{t}_{jn} = t_{jn}c_{jn}.$$

By (i) $\overline{x}_n^{(i)}$ is bounded if $x_n^{(i)}$ is bounded, $i = 1, \dots, m$, by (ii) $\overline{x}_n^{(m_0)}$ is bounded non infinitesimal and by (iii) $\overline{t}_{jn} \to 0$, $j = 1, \dots, \ell$. Moreover $\varphi_i^M(\overline{t}_n)\overline{x}_n^{(i)} = \varphi_i^M(t_n)x_n^{(i)}$, $i = 1, \dots, m$. Applying this process (at most) m_0 times we may find a non infinitesimal bounded sequence $\{\widetilde{u}_n\}_{n\in\mathbb{N}} \in \widetilde{p}^{-1}(Z)$, $\widetilde{u}_n = (\widetilde{x}_n^{(1)}, \dots, \widetilde{x}_n^{(m)}, \widetilde{t}_{1n}, \dots, \widetilde{t}_{\ell n})$ such that

$$\frac{f_k(\varphi_1^M(\widetilde{t}_n)\widetilde{x}_n^{(1)}, \dots, \varphi_1^M(\widetilde{t}_n)\widetilde{x}_n^{(m)})}{\varphi_k^N(\widetilde{t}_n)} = \frac{f_k(\varphi_1^M(t_n)x_n^{(1)}, \dots, \varphi_n^M(t_n)x_n^{(n)})}{\varphi_k^N(t_n)d_n} \to 0.$$

This is because $\{\widetilde{f}(u_n)\}_{n\in\mathbb{N}}$ converges and, by construction of $\widetilde{t}_n$,

$$d_n := \frac{\varphi_k^N(\widetilde{t}_n)}{\varphi_k^N(t_n)} \to +\infty.$$

So we are reduced to find sequences $\{c_{jn}\}_{n\in\mathbb{N}}$, $j = 1, \dots, \ell$ satisfying (i)-(iii), given $u_n = (x_n^{(1)}, \dots, x_n^{(m)}, t_{1n}, \dots, t_{\ell n})$, $t_{jn} > 0$, $j = 1, \dots, \ell$ with $t_{jn} \to 0$ ($j = 1, \dots, \ell$), $\varphi_i^M(t_n)x_n^{(i)} \to 0$ ($i = 1, \dots, m$) and $|x_n^{(m_0)}| \to +\infty$ for $1 \le m_0 \le m$. Assume without loss of generality that $m_0 = 1$. There exists $j_0 \in \{1, \dots, \ell\}$ such that

$$\sup_n \left(\prod_{j \le j_0} t_{jn}^{\alpha_{1j}} |x_n^{(1)}| \right) < +\infty, \qquad \prod_{j<j_0} t_{jn}^{\alpha_{1j}} |x_n^{(1)}| \to +\infty.$$

Let us construct the required sequence as follows:

$$\begin{aligned} c_{jn} &= \frac{1}{t_{jn}\epsilon_n^{\frac{1}{\alpha_{1j}}}} \quad j < j_0, \\ c_{j_0 n} &= \left(\prod_{j<j_0} t_{jn}^{\alpha_{1j}} |x_n^{(1)}| \epsilon_n^{j_0-1} \zeta_n^{j_0-m} \right)^{\frac{1}{\alpha_{1j_0}}}, \\ c_{jn} &= \zeta_n^{\frac{1}{\alpha_{1j}}} \quad j > j_0. \end{aligned}$$

Here $\{\epsilon_n\}_{n\in\mathbb{N}}$ and $\{\zeta_n\}_{n\in\mathbb{N}}$ are sequences, $\epsilon_n, \zeta_n \to +\infty$ slowly enough to satisfy $c_{jn} \to +\infty$ if $j \le j_0$ and $t_{jn}c_{jn} \to 0$ if $j \ge j_0$. One checks easily that (i)-(iii) are satisfied. □

Proposition 3.19. *Let $F \in D^b(k_{Y_{sa}})$.*

(i) There exists a commutative diagram of canonical morphisms

$$\begin{array}{ccc} \omega_{S_{\chi^M}/S_{\chi^N}} \otimes (T_\chi f)^{-1}\nu^{sa}_{\chi^N} F & \longrightarrow & \nu^{sa}_{\chi^M}(\omega_{X/Y}\otimes f^{-1}F) \\ \downarrow & & \downarrow \\ T_\chi f^! \nu^{sa}_{\chi^N} F & \longleftarrow & \nu^{sa}_{\chi^M} f^! F. \end{array}$$

(ii) The above morphisms are isomorphisms on the open set where $T_\chi f$ is smooth.

Proof. (i) The existence of the arrows is done as in [14] Proposition 4.2.5. When (ii) is satisfied the function $\widetilde{f}$ is smooth at any point of the zero section and all the above morphisms become isomorphisms. □

4 Multi-specialization in the complex domain

In this section we extend the notion of asymptotic expansion associated with χ developed in [9]. We refer to [18] for the classical theory of strongly asymptotically developable functions.

4.1 Geometry in the complex domain

Let $X = \mathbb{C}^n$ with coordinates $(z) = (z_1, \ldots, z_n)$, and let $\chi = \{Z_j\}_{j=1}^{\ell}$ be a family of closed complex submanifolds in X. We divide complex coordinates (z) into $(m+1)$-coordinates blocks

$$(z^{(0)}, z^{(1)}, \ldots, z^{(m)}), \tag{4.1}$$

for which each Z_j is assumed to be linearized, that is, there exists the subset K_j in $\{1, 2, \ldots, m\}$ such that each Z_j is defined by the equations $z^{(k)} = 0$ with $k \in K_j$, i.e.,

$$Z_j = \left\{ z^{(k)} = 0 \ (k \in K_j) \right\} \qquad (j = 1, 2, \ldots, \ell). \tag{4.2}$$

Let us consider a matrix $A_\chi = (a_{jk})$ of size $\ell \times m$ whose entries are non-negative rational numbers. Set $L := \operatorname{Rank} A_\chi$. We always assume the condition (2.1), that is,

$$\text{the first } L \times L \text{ submatrix in } A_\chi \text{ is invertible.} \tag{4.3}$$

As in Subsection 2.2, we define the associated m-monomials of $\lambda = (\lambda_1, \ldots, \lambda_\ell)$ by

$$\varphi_k(\lambda) = \lambda_1^{a_{1k}} \lambda_2^{a_{2k}} \cdots \lambda_\ell^{a_{\ell k}} \quad (k = 1, \ldots, m).$$

We also define the action $\mu_j : X \times \mathbb{R}^+ \to X$ $(j = 1, \ldots, \ell)$ by

$$\mu_j(z, t) = (z^{(0)}, t^{a_{j1}} z^{(1)}, t^{a_{j2}} z^{(2)}, \ldots, t^{a_{jm}} z^{(m)}) \tag{4.4}$$

and the action $\mu : X \times (\mathbb{R}^+)^\ell \to X$ by

$$\mu(z, \lambda) = (z^{(0)}, \varphi_1(\lambda)z^{(1)}, \varphi_2(\lambda)z^{(2)}, \ldots, \varphi_m(\lambda)z^{(m)}). \tag{4.5}$$

For these actions, we assume Condition A2. introduced in Subsection 1.4, that is,

$$\text{each } Z_j \text{ coincides with the set of fixed points of the action } \mu_j \text{ on } X. \tag{4.6}$$

Note that this condition is equivalently saying that A_χ satisfies the condition

$$a_{jk} \neq 0 \text{ if and only if } k \in K_j \qquad (j = 1, 2, \ldots, \ell). \tag{4.7}$$

According to the system of coordinates $(z^{(0)}, \ldots, z^{(m)})$, the space X can be identified with

$$\mathbb{C}^{n_0}_{z^{(0)}} \times \mathbb{C}^{n_1}_{z^{(1)}} \times \cdots \times \mathbb{C}^{n_m}_{z^{(m)}},$$

where n_k denotes the number of coordinate variables in the coordinates block $z^{(k)}$.

Remark 4.1. *Through the section, for a complex vector $a = (a_1, \ldots, a_k) \in \mathbb{C}^k$, its norm is defined by*

$$|a| := \max\{|a_1|, |a_2|, \ldots, |a_k|\}. \tag{4.8}$$

Let p be a point $(z^{(0)}; \zeta) = (0; \zeta^{(1)}, \ldots, \zeta^{(m)}) \in S_\chi$ and set

$$J_Z(\zeta) = \{k \in \{1, \ldots, m\};\ \zeta^{(k)} = 0\} \tag{4.9}$$

as usual. Note that we consider the problem near the origin, i.e., $z^{(0)} = 0$. Further, for the point $p \in S_\chi$, we assume the condition (2.3), that is,

$$\text{if the } k\text{-th column of } A_\chi \text{ is a zero vector, then } \zeta^{(k)} = 0. \tag{4.10}$$

We denote by $V(\zeta)$ the set of $\mathbb{R}^+$-cones in the form

$$\mathbb{C}^{n_0} \times V_1 \times \cdots \times V_m \subset X, \tag{4.11}$$

where each V_k satisfies the following conditions:

- If $\zeta^{(k)} \neq 0$, then V_k is an open convex proper cone in $\mathbb{C}^{n_k}$ containing the point $\zeta^{(k)}$.
- If $\zeta^{(k)} = 0$, then $V_k = \mathbb{C}^{n_k}$.

In what follows, we constantly use notions introduced in Section 2. Remember that the semigroup $\mathcal{G}$ was defined in Subsection 2.2, and that F^q (resp. G) is the finite subset of $\mathcal{H}_\tau \times \mathbb{R}_{\geq 0}$ (resp. $\mathcal{H}_{\tau,\lambda} \times \mathbb{R}_{\geq 0}$) defined in the same subsection, where we adopt the definition of G given in Remark 2.10. That is,

$$F^0 := (\mathcal{L}^\lambda_\ell \circ \cdots \circ \mathcal{L}^\lambda_{L+1})(G),$$
$$F^q := (\mathcal{L}_{k_q} \circ \cdots \circ \mathcal{L}_{k_1})(F^0),$$

where $\{k_1, \ldots, k_q\} = J_Z(\zeta) \cap \{1, 2, \ldots, L\}$ $(1 \leq k_1 < \cdots < k_q \leq L)$. Note that, by Proposition 2.28, the radical of $[F^q]$ is $\mathcal{G}$.

Remark 4.2. *The most important and interesting cases are those where the corresponding action μ_X is non-degenerate and p is outside the fixed points. In this case, by adopting the coordinates blocks described in Corollary 2.5, we have $q = 0$ and $L = \operatorname{Rank} A_X = \ell$, and hence,*

$$F^q = F^0 = G. \tag{4.12}$$

Set

$$|z^{(*)}| = (|z^{(1)}|, \dots, |z^{(L)}|, \dots, |z^{(m)}|) \tag{4.13}$$

and

$$|z^{(*)}|' = (|z^{(1)}|, \dots, |z^{(L)}|). \tag{4.14}$$

Let H be a finite subset in $\mathcal{H}_\tau \times \mathbb{R}_{\geq 0}$ satisfying that $[H]$ and $\mathcal{G}$ defined in (2.29) are equivalent, and let $\{\epsilon\}$ be the set of positive real numbers consisting of $\epsilon_0 > 0$, $\epsilon_{(f,v),+} > 0$ and $\epsilon_{(f,v),-} > 0$ for each $(f, v) \in H$. Then, for $V \in V(\zeta)$, we define the open subset

$$S_H(V, \{\epsilon\}) := \left\{ z \in X;\ \begin{array}{c} z \in V,\ |z^{(0)}| < \epsilon_0, \\ v - \epsilon_{(f,v),-} < f\left(|z^{(*)}|\right) < v + \epsilon_{(f,v),+} \\ \text{for any } (f, v) \in H \end{array} \right\}, \tag{4.15}$$

which is often called a *multi-cone* or *multi-sector*. Since $[H]$ and $\mathcal{G}$ are equivalent and $(\tau_k^N, 0) \in \mathcal{G}$ $(k = 1, 2, \dots, m)$ holds for some $N \in \mathbb{N}$, any open neighborhood of the origin contains the set $S_H(V, \{\epsilon\})$ if the positive real numbers in $\{\epsilon\}$ are sufficiently small.

Furthermore, for finite subsets H and H' in $\mathcal{H}_\tau \times \mathbb{R}_{\geq 0}$ such that both the $[H]$ and $[H']$ are equivalent, the families $\{S_H(V, \{\epsilon\})\}_{\{\epsilon\}}$ and $\{S_{H'}(V, \{\epsilon'\})\}_{\{\epsilon'\}}$ are equivalent with respect to inclusion of sets. That is, for any $S_H(V, \{\epsilon\})$ (resp. $S_{H'}(V, \{\epsilon'\})$), there exists $\{\epsilon'\}$ (resp. $\{\epsilon\}$) such that

$$S_H(V, \{\epsilon\}) \subset S_{H'}(V, \{\epsilon'\}) \quad (\text{resp. } S_{H'}(V, \{\epsilon'\}) \subset S_H(V, \{\epsilon\}))$$

holds.

Lemma 4.3. *Let $\mathcal{G}'$ be a semigroup in $\mathcal{H}_\tau \times \mathbb{R}_{\geq 0}$ which is equivalent to $\mathcal{G}$. Then, the family $\{S_H(V, \{\epsilon\})\}_{H, \{\epsilon\}}$ and the one $\{S_{H'}(V, \{\epsilon'\})\}_{H', \{\epsilon'\}}$ are exactly same where the former H runs through the finite subset of $\mathcal{G}$ with $[H]$ and $\mathcal{G}$ being equivalent and the latter H' runs through the finite subset of $\mathcal{G}'$ with $[H']$ and $\mathcal{G}'$ being equivalent. That is, for any $S_H(V, \{\epsilon\})$ (resp. $S_{H'}(V, \{\epsilon'\})$), there exists a finite subset $H' \subset \mathcal{G}'$ (resp. $H \subset \mathcal{G}$) and $\{\epsilon'\}$ (resp. $\{\epsilon\}$) such that $S_H(V, \{\epsilon\}) = S_{H'}(V, \{\epsilon'\})$ holds.*

Proof. Let H be a finite subset in $\mathcal{G}$ satisfying that $[H]$ and $\mathcal{G}$ are equivalent. As $\mathcal{G}'$ is equivalent to $\mathcal{G}$, there exists a natural number N such that $\langle H\rangle_N \subset \mathcal{G}'$ holds, where $\langle H\rangle_N$ is the finite subset $\{(f, v)^N;\ (f, v) \in H\}$. Note that $[\langle H\rangle_N]$ is equivalent to $\mathcal{G}'$.

On the other hand, it follows from the definition that, for any $N \in \mathbb{N}$ and $\{\epsilon\}$, there exists a suitable $\{\epsilon'\}$ for which we have

$$S_H(V,\ \{\epsilon\}) = S_{\langle H\rangle_N}(V,\ \{\epsilon'\}).$$

Hence the claim of the lemma follows from this. □

Remark 4.4. *When $Q(H) = H$ holds, for any $(f, v) \in H$ with $v \neq 0$, we have $(1/f, 1/v) \in H$. Hence, by suitably choosing $\{\epsilon\}$ again, the multi-cone defined by (4.15) coincides with the subset*

$$\left\{ z \in X;\ \begin{matrix} z \in V,\ |z^{(0)}| < \epsilon_0, \\ f\left(|z^{(*)}|\right) < v + \epsilon_{(f,v),+} \text{ for any } (f,\ v) \in H \end{matrix} \right\}. \tag{4.16}$$

Remark 4.5. *When $H = F^q \subset \mathcal{G}$ and all the positive real numbers in $\{\epsilon\}$ are the same $\epsilon > 0$, we write $S(V,\ \epsilon)$ instead of $S_H(V,\ \{\epsilon\})$.*

Example 4.6. *Let $X = \mathbb{C}^n$ with coordinates blocks $(z^{(0)},\ z^{(1)},\ z^{(2)},\ z^{(3)})$. The submanifolds Z_j $(j = 1, 2, 3)$ are defined by*

$$Z_1 = \{z^{(2)} = z^{(3)} = 0\},$$
$$Z_2 = \{z^{(1)} = z^{(3)} = 0\},$$
$$Z_3 = \{z^{(3)} = 0\}.$$

Let us consider the 3×3 matrix A_χ associated with the family $\chi = \{Z_1,\ Z_2,\ Z_3\}$. Then clearly we have $\operatorname{Rank} A_\chi = 3$ and χ is of normal type. In this case, the polynomials $\varphi_j(\lambda)$ are given by

$$\varphi_1(\lambda) = \lambda_2,\ \varphi_2(\lambda) = \lambda_1,\ \varphi_3(\lambda) = \lambda_1\lambda_2\lambda_3,$$

and hence, we have

$$\varphi_1^{-1}(\tau) = \tau_1,\ \varphi_2^{-1}(\tau) = \tau_2,\ \varphi_3^{-1}(\tau) = \frac{\tau_3}{\tau_1\tau_2}.$$

Let $p = (0; \zeta) = (0; \zeta^{(1)}, \zeta^{(2)}, \zeta^{(3)}) \in S_\chi$ be a point outside fixed points. Since G is given by

$$\{(\varphi_1^{-1},\ 0),\ (\varphi_2^{-1},\ 0),\ (\varphi_3^{-1},\ 0)\}$$

and $F^q = G$ holds in this case, the cone $S(V, \epsilon)$ is defined by

$$\left\{ (z^{(0)},\ z^{(1)},\ z^{(2)},\ z^{(3)}) \in X;\ \begin{matrix} z^{(k)} \in V_k\ (k = 1, 2, 3),\ |z^{(j)}| < \epsilon\ (j = 0, 1, 2, 3), \\ |z^{(3)}| < \epsilon|z^{(1)}||z^{(2)}| \end{matrix} \right\},$$

where V_j $(j = 1, 2, 3)$ is a proper open cone in $\mathbb{C}^{n_j}$ containing the vector $\zeta^{(j)}$.

Example 4.7. *Let $X = \mathbb{C}^n$ with coordinates blocks $(z^{(0)}, z^{(1)}, z^{(2)}, z^{(3)}, z^{(4)})$. The submanifolds Z_j $(j = 1, 2, 3, 4)$ are defined by*

$$\begin{aligned} Z_1 &= \{z^{(1)} = z^{(4)} = 0\}, \\ Z_2 &= \{z^{(2)} = z^{(4)} = 0\}, \\ Z_3 &= \{z^{(3)} = z^{(4)} = 0\}, \\ Z_4 &= \{z^{(1)} = z^{(2)} = z^{(3)} = z^{(4)} = 0\}. \end{aligned}$$

Let us consider the 4×4 matrix A_χ associated with the family $\chi = \{Z_1, Z_2, Z_3, Z_4\}$. Then $\operatorname{Rank} A_\chi = 4$ holds and χ is of normal type. In this case, the polynomials $\varphi_j(\lambda)$ are given by

$$\varphi_1(\lambda) = \lambda_1\lambda_4,\ \varphi_2(\lambda) = \lambda_2\lambda_4,\ \varphi_3(\lambda) = \lambda_3\lambda_4,\ \varphi_4(\lambda) = \lambda_1\lambda_2\lambda_3\lambda_4,$$

and hence

$$\varphi_1^{-1}(\tau) = \sqrt{\frac{\tau_1\tau_4}{\tau_2\tau_3}},\ \varphi_2^{-1}(\tau) = \sqrt{\frac{\tau_2\tau_4}{\tau_1\tau_3}},\ \varphi_3^{-1}(\tau) = \sqrt{\frac{\tau_3\tau_4}{\tau_1\tau_2}},\ \varphi_4^{-1}(\tau) = \sqrt{\frac{\tau_1\tau_2\tau_3}{\tau_4}}.$$

Let $p = (0; \zeta) = (0; \zeta^{(1)}, \zeta^{(2)}, \zeta^{(3)}, \zeta^{(4)}) \in S_\chi$ be a point outside fixed points. By the same reasoning as that in the previous example, the cone $S(V, \epsilon)$ is defined by

$$\left\{ (z^{(0)}, z^{(1)}, z^{(2)}, z^{(3)}, z^{(4)}) \in X;\ \begin{aligned} &z^{(k)} \in V_k\ (k = 1, 2, 3, 4),\ |z^{(0)}| < \epsilon, \\ &|z^{(3)}||z^{(4)}| < \epsilon^2|z^{(1)}||z^{(2)}|, \\ &|z^{(2)}||z^{(4)}| < \epsilon^2|z^{(1)}||z^{(3)}|, \\ &|z^{(1)}||z^{(4)}| < \epsilon^2|z^{(2)}||z^{(3)}|, \\ &|z^{(1)}||z^{(2)}||z^{(3)}| < \epsilon^2|z^{(4)}| \end{aligned} \right\},$$

where V_j $(j = 1, 2, 3, 4)$ is a proper open cone in $\mathbb{C}^{n_j}$ containing the vector $\zeta^{(j)}$.

Lemma 4.8. *Assume the positive real numbers in $\{\epsilon\}$ to be sufficiently small. Then $S_H(V, \{\epsilon\})$ is 1-regular, that is, there exists a constant $C > 0$ satisfying that, for any points p and q in $S_H(V, \{\epsilon\})$, there exists a rectifiable curve in $S_H(V, \{\epsilon\})$ which joins p and q and whose length is less than or equal to $C|p - q|$.*

Proof. We first show the fact that the subset D in $\mathbb{R}^m$ defined by

$$\left\{ (\tau_1, \ldots, \tau_m) \in \mathbb{R}^m;\ \begin{aligned} &\tau_k > 0\ (k = 1, \ldots, m), \\ &v - \epsilon_{(f,v),-} < f(\tau) < v + \epsilon_{(f,v),+} \\ &\text{for any } (f,\, v) \in H \end{aligned} \right\}$$

is 1-regular. Take points

$$a = (a_1,\ \ldots,\ a_m) \in D$$

and

$$b = (b_1, \ldots, b_m) \in D.$$

Consider the closed curve parameterized by θ $(0 \le \theta \le 1)$

$$(a_1^\theta b_1^{(1-\theta)}, \ldots, a_m^\theta b_m^{(1-\theta)}).$$

Since $f \in H$ is a monomial of the τ variables, the curve is contained in D. Furthermore, as $a_k^\theta b_k^{(1-\theta)}$ is a monotonic function of θ, the length of the curve is dominated by

$$\sum_{k=1}^m |a_k - b_k| \le m|a-b|.$$

Hence D is 1-regular. Now define

$$Y = \{z \in S_H(V, \{\epsilon\});\ |z^{(1)}||z^{(2)}|\cdots|z^{(m)}| = 0\}.$$

Since $S_H(V, \{\epsilon\}) \setminus Y$ is dense in $S_H(V, \{\epsilon\})$, we may assume p and q in the lemma to be outside Y. Let

$$p = (\hat{z}^{(0)}, \hat{z}^{(1)}, \ldots, \hat{z}^{(m)}) \in S_H(V, \{\epsilon\}) \setminus Y$$

and

$$q = (\hat{w}^{(0)}, \hat{w}^{(1)}, \ldots, \hat{w}^{(m)}) \in S_H(V, \{\epsilon\}) \setminus Y.$$

Then we take points

$$p_1 = (\hat{w}^{(0)}, \hat{z}^{(1)}, \ldots, \hat{z}^{(m)}) \in S_H(V, \epsilon) \setminus Y$$

and

$$p_2 = \left(\hat{w}^{(0)}, \frac{|\hat{w}^{(1)}|}{|\hat{z}^{(1)}|}\hat{z}^{(1)}, \ldots, \frac{|\hat{w}^{(m)}|}{|\hat{z}^{(m)}|}\hat{z}^{(m)}\right) \in S_H(V, \{\epsilon\}) \setminus Y.$$

We join these points successively: First p and p_1 are joined by the straight segment. Let P be the product of the spherical surfaces defined by

$$\{(z^{(0)}, z^{(1)}, \ldots, z^{(m)});\ |z^{(k)}| = |\hat{w}^{(k)}|\ \ (k = 1, \ldots, m)\}.$$

Then the point p_2 and q are connected by the shortest arc in the surface P. Since D is 1-regular, there exists a curve $(\tau_1(\theta), \ldots, \tau_m(\theta))$ $(0 \le \theta \le 1)$ in D which joins the points $a = (|z^{(1)}|, \ldots, |z^{(m)}|)$ and $b = (|w^{(1)}|, \ldots, |w^{(m)}|)$ and whose length is bounded by $m|a-b|$. Then p_1 and p_2 are jointed by the closed curve $(0 \le \theta \le 1)$

$$\left(\hat{w}^{(0)}, \frac{\tau_1(\theta)}{|\hat{z}^{(1)}|}\hat{z}^{(1)}, \ldots, \frac{\tau_m(\theta)}{|\hat{z}^{(m)}|}\hat{z}^{(m)}\right).$$

Clearly these curves are contained in $S_H(V, \{\epsilon\})$ and the total length of these curves is bounded by a constant multiple of $|p-q|$. □

By the same method as above, we can also prove the following corollary:

Corollary 4.9. *Assume the positive real numbers in $\{\epsilon\}$ to be sufficiently small. Then, for any point $p \in X$, there exists a family $\{U_\kappa\}_{\kappa>0}$ of open subanalytic neighborhoods of p such that $U_\kappa \cap S_H(V, \{\epsilon\})$ is contractible. In particular, $S_H(V, \{\epsilon\})$ is locally cohomologically trivial.*

Proof. We may assume I_0 to be empty. Then every point has a fundamental open neighborhood in the form

$$\{(z^{(1)}, \ldots, z^{(m)});\ a_k < |z^{(k)}| < b_k,\ z^{(k)} \in T_k\ (k = 1, \ldots, m)\},$$

where a_k, b_k and T_k satisfy the one of the following conditions:

1. $0 < a_k < b_k$ and T_k is an open convex proper cone in $\mathbb{C}^{n_k}_{z^{(k)}}$.
2. $a_k < 0 < b_k$ and $T_k = \mathbb{C}^{n_k}_{z^{(k)}}$.

Let U be such an open subset.

We first suppose $\zeta^{(k)} \neq 0$ for any $k \in \{1, 2, \ldots, m\}$ and $U \cap S_H(V, \{\epsilon\})$ is non-empty. Take a point q in $U \cap S_H(V, \{\epsilon\})$ and fix it. Then, by the same argument as that for the proof of the previous lemma, every point $p \in U \cap S_H(V, \{\epsilon\})$ can be joined with q by an explicitly specified path in $U \cap S_H(V, \{\epsilon\})$. Hence the family of these paths give a contraction map.

Now let us consider the general case. Set

$$K_Z := \{k \in \{1, \ldots, m\};\ \zeta^{(k)} = 0,\ a_k < 0\}.$$

For any $p = (z_*^{(1)}, \ldots, z_*^{(m)}) \in U \cap S_H(V, \{\epsilon\})$, we determine the point $\tilde{p} = (\tilde{z}_*^{(1)}, \ldots, \tilde{z}_*^{(m)})$ by $\tilde{z}_*^{(k)} = 0$ if $k \in K_Z$ and $\tilde{z}_*^{(k)} = z_*^{(k)}$ if $k \notin K_Z$. Then the straight segment joining p and $\tilde{p}$ is completely contained in $U \cap S_H(V, \{\epsilon\})$ because, for $(f(\tau), v) \in H$, the variable τ_k with $k \in K_Z$ appears only in the numerator of f, and further, $v = 0$ when τ_k appears in the numerator of f. Therefore we first change every point p in $U \cap S_H(V, \{\epsilon\})$ to $\tilde{p}$, and then, apply the previous argument to the only coordinates $z^{(k)}$ $(k \notin K_Z)$, i.e., $z^{(k)}$ with $k \in K_Z$ is fixed to be the origin in $\mathbb{C}^{n_k}$. □

Hereafter we always assume that the positive real numbers in $\{\epsilon\}$ are sufficiently small so that the above lemma and corollary hold for $S_H(V, \{\epsilon\})$.

We denote by $\widehat{\mathcal{P}}_\ell$ the set of all the subsets of $\{1, 2, \ldots, \ell\}$ except for the empty set. For any $J \in \widehat{\mathcal{P}}_\ell$, we set

$$K_J := \bigcup_{j \in J} K_j, \qquad Z_J := \bigcap_{j \in J} Z_j. \tag{4.17}$$

For any subset $K \subset \{1, 2, \ldots, m\}$, we denote by $z^{(K)}$ the set of the coordinates blocks $z^{(k)}$'s with $k \in K$. We also set

$$\complement K := \{0, 1, \ldots, m\} \setminus K.$$

Note that the coordinates of Z_J are given by $z^{(\complement K_J)}$. Let π_J denote the canonical projection from X to Z_J defined by $z \to z^{(\complement K_J)}$.

Lemma 4.10. *Let $V \in V(\zeta)$, and let $\{\epsilon\}$ be the set of positive real numbers. Then we have the followings:*

1. *For a $\lambda = (\lambda_1, \ldots, \lambda_\ell)$ with $0 < \lambda_j \leq 1$ $(j = 1, 2, \ldots, \ell)$, we have*

$$\mu(S_H(V, \{\epsilon\}), \lambda) \subset S_H(V, \{\epsilon\}).$$

2. *For a $J \in \widehat{\mathcal{P}}_\ell$, we have*

$$\overline{\pi_J(S_H(V, \{\epsilon\}))} = Z_J \cap \overline{S_H(V, \{\epsilon\})}.$$

Proof. The first part is already proved in Section 2, Corollary 2.36.

Let us show the second claim: Set $S := S_H(V, \{\epsilon\})$ for short. Clearly

$$\overline{\pi_J(S)} \supset \pi_J\left(\overline{S}\right) \supset \overline{S} \cap Z_J$$

hold. Now we will show the converse inclusion. For $\lambda = (\lambda_1, \ldots, \lambda_\ell)$ and $J = \{j_1, \ldots, j_r\}$ $(j_1 < \cdots < j_r)$, set $\lambda_J := (\lambda_{j_1}, \ldots, \lambda_{j_r})$ and define $\mu_J : X \times (\mathbb{R}^+)^{\#J} \to X$ by

$$\mu_J(z, \lambda_J) = \mu_{j_1}(\mu_{j_2}(\ldots(\mu_{j_r}(z, \lambda_{j_r}), \ldots), \lambda_{j_2}), \lambda_{j_1}).$$

Set

$$\tilde{\mu}_J(z, t) = \mu_J(z, t, \ldots, t) \quad (t \in \mathbb{R}).$$

Suppose $z_\infty \in \overline{\pi_J(S)}$. Then we can find a sequence $\{z_m\}_{m=1}^\infty$ such that $z_m \in S$ and $\pi_J(z_m) \to z_\infty$. Since $\pi_J(z_m) = \tilde{\mu}_J(z_m, 0)$ holds, the sequence $\tilde{z}_m := \tilde{\mu}_J(z_m, 1/m)$ tends to z_∞. By the claim 1. we have $\tilde{z}_m \in S$, from which $z_\infty \in \overline{S}$ follows. This completes the proof. □

For a $J \in \widehat{\mathcal{P}}_\ell$ and for an $S := S_H(V, \{\epsilon\})$, we set $S_J := \pi_J(S)$, and by the above lemma, we have

$$\overline{S_J} = \overline{S} \cap Z_J. \tag{4.18}$$

S_J enjoy several good properties as those in S. Let $\mu|_{Z_J}$ (resp. $\mu_j|_{Z_J}$) be the restriction of the action μ (resp. μ_j) to Z_J.

Proposition 4.11. *Assume $Q(H) = H$. Then, for $S := S_H(V, \{\epsilon\})$ and $J \in \widehat{\mathcal{P}}_\ell$, the subset S_J is a multi-cone in Z_J with respect the action $\mu|_{Z_J}$. In particular, we have:*

1. *S_J is stable under contraction by the action $\mu_j|_{Z_J}$, that is, $\mu_j|_{Z_J}(z,t) \in S_J$ if $z \in S_J$ and $0 < t \le 1$.*
2. *S_J is 1-regular and locally cohomologically trivial if $\{\epsilon\}$ consists of sufficiently small positive numbers.*

Proof. It follows from Proposition 2.13 that $\mathcal{G}$ and $\widehat{\mathcal{G}}$ are equivalent. Hence, by Lemma 4.3, we can assume that $H \subset \widehat{\mathcal{G}}$ and $[H]$ and $\widehat{\mathcal{G}}$ are equivalent. Furthermore, since $Q(H) = H$ holds, we also assume that $S_H(V, \{\epsilon\})$ is defined by the special form (4.16). Suppose that K_J is non-empty and

$$K_J = \{m' + 1,\, m' + 2,\, \ldots,\, m\}$$

for some $0 \le m' < m$. Set

$$\tau_J = (\tau_1, \ldots, \tau_{m'})$$

and denote by $\mathcal{H}_{\tau_J}$ the set of monomials of rational powers of the variables τ_J with coefficients 1.

Recall that $\widehat{\mathcal{G}}$ is given by

$$(\mathcal{H}_\tau \times \mathbb{R}_{\ge 0}) \bigcap [\,\widehat{G}\,],$$

where $\widehat{G}$ is

$$Q\left(\left\{\left(\frac{\tau_1}{\varphi_1(\lambda)}, |\zeta^{(1)}|\right), \cdots, \left(\frac{\tau_m}{\varphi_m(\lambda)}, |\zeta^{(m)}|\right), (\lambda_1, 0), \cdots, (\lambda_\ell, 0)\right\}\right).$$

The corresponding semigroup $\widehat{\mathcal{G}}_J$ with respect to the action $\mu|_{Z_J}$ in Z_J is given by

$$(\mathcal{H}_{\tau_J} \times \mathbb{R}_{\ge 0}) \bigcap [\,\widehat{G}_J\,],$$

where $\widehat{G}_J$ is

$$Q\left(\left\{\left(\frac{\tau_1}{\varphi_1(\lambda)}, |\zeta^{(1)}|\right), \cdots, \left(\frac{\tau_{m'}}{\varphi_{m'}(\lambda)}, |\zeta^{(m')}|\right), (\lambda_1, 0), \cdots, (\lambda_\ell, 0)\right\}\right).$$

By noticing the fact that the variable τ_k appears only in the term $\tau_k/\varphi_k(\lambda)$, we can easily see

$$(\mathcal{H}_{\tau_J} \times \mathbb{R}_{\ge 0}) \bigcap \widehat{\mathcal{G}} = \widehat{\mathcal{G}}_J.$$

Recall the operation $\mathcal{L}_k$ defined in Subsection 2.3. Let us define the slightly modified operation $\widehat{\mathcal{L}}_k$ which consists of the only F1 and F3 in the definition of $\mathcal{L}_k$. Then, as the same reasoning as that for the claim 2. in Proposition 2.28, the finite subset defined by

$$H_J := (\widehat{\mathcal{L}}_m \circ \cdots \circ \widehat{\mathcal{L}}_{m'+1})(H)$$

is the radical of $\widehat{\mathcal{G}}_J$. Now, by repeated application of Lemma 4.12 below, we can conclude that $S_J = \pi_J(S)$ is a multi-cone $S_{H_J}(\pi_J(V), \{\epsilon'\})$ in Z_J with suitably chosen $\{\epsilon'\}$. This completes the proof. □

Before stating the lemma, we need some preparations: Let T be the product space $T_1 \times T_2 \times \cdots \times T_m$ with the coordinates $\tau = (\tau_1, \ldots, \tau_m)$ where each T_k is either $\mathbb{R}^+$ or $\mathbb{R}_{\geq 0}$. Set $T' := T_2 \times \cdots \times T_m$ with the coordinates $\tau' = (\tau_2, \ldots, \tau_m)$ and $\pi_1 : T \to T'$ is the canonical projection $\tau \to \tau'$.

Let $\{g_s\}_{s \in \Lambda}$ be a finite subset in $\mathcal{H}_\tau$. We assume that each g_s is well defined on T, that is, $\nu_k(g_s) \geq 0$ for any $1 \leq k \leq m$ with $T_k = \mathbb{R}_{\geq 0}$. Let $\{a_s\}_{s \in \Lambda}$ be a set of positive real numbers and define the subset

$$W := \{\tau \in T;\ g_s(\tau) < a_s\ (s \in \Lambda)\}.$$

Then, under this situation, we have the lemma below:

Lemma 4.12. *The subset $\pi_1(W) \subset T'$ is defined by the following family of inequalities:*

$$g_s < a_s \tag{4.19}$$

for any g_s $(s \in \Lambda)$ with $\nu_1(g_s) = 0$ and

$$g_s^a g_{s'}^b < a_s^a a_{s'}^b \tag{4.20}$$

for any pair g_s and $g_{s'}$ $(s, s' \in \Lambda)$ with $\nu_1(g_s) < 0$ and $\nu_1(g_{s'}) > 0$, and natural numbers a and b are taken to be prime to each other with $a|\nu_1(g_s)| = b|\nu_1(g_{s'})|$.

Remark 4.13. *The inequalities* (4.19) *and* (4.20) *are counterparts of F1 and F3 of the operation $\mathcal{L}_1$, respectively.*

Proof. If $T_1 = \mathbb{R}_{\geq 0}$, then the lemma trivially holds. Hence we assume $T_1 = \mathbb{R}^+$ and, for simplicity, we may also assume $\nu_1(g_s) \neq 0$ for any $s \in \Lambda$. In this case, $\tau' \in \pi_1(W)$ if and only if we have

$$\max_{s \in \Lambda,\, \nu_1(g_s) < 0} \mathrm{sol}(g_s, a_s)(\tau') \ < \min_{s \in \Lambda,\, \nu_1(g_s) > 0} \mathrm{sol}(g_s, a_s)(\tau'),$$

where $\mathrm{sol}(g_s, a_s)(\tau')$ is the solution of the equation $g_s(\tau_1, \tau') = a_s$ when we regarded it as an equation of τ_1. The latter condition is clearly equivalent to (4.20) for any pair g_s and $g_{s'}$ $(s, s' \in \Lambda)$ with $\nu_1(g_s) < 0$ and $\nu_1(g_{s'}) > 0$. □

Remark 4.14. *Some $\mu_j|_{Z_J}$ may be the identity action on Z_J. Since the geometrical situation remains unchanged if the corresponding deformation parameter t_j is removed. Hence we may consider the situation where such an action is ignored and the corresponding deformation parameter is removed. Furthermore, if there is a pair μ_j and $\mu_{j'}$ $(j \neq j')$ which gives the same induced actions $\mu_j|_{Z_J} = \mu_{j'}|_{Z_J}$, we can also eliminate such a duplication, for example, by removing $\mu_j|_{Z_J}$ and the corresponding deformation parameter t_j.*

For a multi-cone, we can introduce the notion of a proper sub-multi-cone as follows: Recall that, for $V = \mathbb{C}^{n_0} \times V_1 \times \cdots \times V_m$ and $V' = \mathbb{C}^{n_0} \times V_1' \times \cdots \times V_m'$ in $V(\zeta)$, we say that V is *properly contained* in V' if $\overline{V_k} \setminus \{0\} \subset V_k'$ holds for any $k = 1, 2, \ldots, m$.

Definition 4.15. *We say that $S = S_H(V, \{\epsilon\})$ is properly contained in $S' = S_H(V', \{\epsilon'\})$ if V is properly contained in V' and if the positive real number in $\{\epsilon\}$ is strictly smaller than the corresponding one in $\{\epsilon'\}$, i.e., $\epsilon_0 < \epsilon_0'$, $\epsilon_{(f,v),-} < \epsilon'_{(f,v),-}$ and $\epsilon_{(f,v),+} < \epsilon'_{(f,v),+}$ hold for any $(f, v) \in H$.*

4.2 Multi-specialization and asymptotic expansions

We give a functorial construction of the sheaf of multi-asymptotically developable functions using the sheaf of Whitney $\mathcal{C}^\infty$-functions of [16]. Let X be a real analytic manifold ($X = \mathbb{R}^n$ in the local model). As usual, given $F \in D^b(\mathbb{C}_X)$ we set $D'F = R\mathcal{H}om(F, \mathbb{C}_X)$. Remember that an open subset U of X is locally cohomologically trivial (l.c.t. for short) if $D'\mathbb{C}_U \simeq \mathbb{C}_{\overline{U}}$. Let $\overset{\mathrm{w}}{\otimes}$ denote the Whitney tensor product of [15].

Definition 4.16. *Let $F \in \mathrm{Mod}_{\mathbb{R}\text{-c}}(\mathbb{C}_X)$ and let $U \in \mathrm{Op}(X_{sa})$. We define the presheaf $\mathcal{C}^{\infty,\mathrm{w}}_{X|F}$ as follows:*

$$U \mapsto \Gamma(X; H^0 D'\mathbb{C}_U \otimes F \overset{\mathrm{w}}{\otimes} \mathcal{C}^\infty_X).$$

Let $U, V \in \mathrm{Op}(X_{sa})$, and consider the exact sequence

$$0 \to \mathbb{C}_{U\cap V} \to \mathbb{C}_U \oplus \mathbb{C}_V \to \mathbb{C}_{U\cup V} \to 0,$$

applying the functor $\mathcal{H}om(\cdot, \mathbb{C}_X) = H^0 D'(\cdot)$ we obtain

$$0 \to H^0 D'\mathbb{C}_{U\cup V} \to H^0 D'\mathbb{C}_U \oplus H^0 D'\mathbb{C}_V \to H^0 D'\mathbb{C}_{U\cap V},$$

applying the exact functors $\cdot \otimes F$, $\cdot \overset{\mathrm{w}}{\otimes} \mathcal{C}^\infty_X$ and taking global sections we obtain

$$0 \to \mathcal{C}^{\infty,\mathrm{w}}_{X|F}(U \cup V) \to \mathcal{C}^{\infty,\mathrm{w}}_{X|F}(U) \oplus \mathcal{C}^{\infty,\mathrm{w}}_{X|F}(V) \to \mathcal{C}^{\infty,\mathrm{w}}_{X|F}(U \cap V).$$

This implies that $\mathcal{C}^{\infty,\mathrm{w}}_{X|F}$ is a sheaf on X_{sa}. Moreover if $U \in \mathrm{Op}(X_{sa})$ is l.c.t., the morphism $\Gamma(X; \mathcal{C}^{\infty,\mathrm{w}}_{X|F}) \to \Gamma(U; \mathcal{C}^{\infty,\mathrm{w}}_{X|F})$ is surjective and $R\Gamma(U; \mathcal{C}^{\infty,\mathrm{w}}_{X|F})$ is concentrated in degree zero. Let $0 \to F \to G \to H \to 0$ be an exact sequence in $\mathrm{Mod}_{\mathbb{R}\text{-c}}(\mathbb{C}_X)$, we obtain an exact sequence in $\mathrm{Mod}(\mathbb{C}_{X_{sa}})$

$$0 \to \mathcal{C}^{\infty,\mathrm{w}}_{X|F} \to \mathcal{C}^{\infty,\mathrm{w}}_{X|G} \to \mathcal{C}^{\infty,\mathrm{w}}_{X|H} \to 0. \tag{4.21}$$

We can easily extend the sheaf $\mathcal{C}^{\infty,\mathrm{w}}_{X|F}$ to the case of $F \in D^b_{\mathbb{R}\text{-c}}(\mathbb{C}_X)$, taking a finite resolution of F consisting of locally finite sums $\oplus \mathbb{C}_V$ with V l.c.t. in $\mathrm{Op}^c(X_{sa})$. In fact, the sheaves $\mathcal{C}^{\infty,\mathrm{w}}_{X|\oplus\mathbb{C}_V}$ form a complex quasi-isomorphic to $\mathcal{C}^{\infty,\mathrm{w}}_{X|F}$ consisting of acyclic objects with respect to $\Gamma(U;\cdot)$, where U is l.c.t. in $\mathrm{Op}^c(X_{sa})$.

As in the case of Whitney $\mathcal{C}^\infty$-functions one can prove that, if $G \in D^b_{\mathbb{R}\text{-c}}(\mathbb{C}_X)$ one has

$$\rho^{-1}R\mathcal{H}om(G,\mathcal{C}^{\infty,\mathrm{w}}_{F|X}) \simeq D'G \otimes F \overset{\mathrm{w}}{\otimes} \mathcal{C}^\infty_X .$$

Example 4.17. *Setting $F = \mathbb{C}_X$ we obtain the sheaf of Whitney $\mathcal{C}^\infty$-functions. Indeed, let U be a l.c.t. subanalytic open subset of X. Then $\Gamma(U;\mathcal{C}^{\infty,\mathrm{w}}_{X_\mathbb{R}}) \simeq \Gamma(X;\mathbb{C}_{\overline{U}} \overset{\mathrm{w}}{\otimes} \mathcal{C}^\infty_{X_\mathbb{R}})$ is nothing but the set of $\mathcal{C}^\infty$-Whitney jets on $\overline{U}$. Let Z be a closed subanalytic subset of X. Similarly one checks that $\mathcal{C}^{\infty,\mathrm{w}}_{X|\mathbb{C}_{X\setminus Z}}$ is the sheaf of Whitney $\mathcal{C}^\infty$-functions vanishing on Z with all their derivatives.*

Notation 4.18. *Let S be a locally closed subanalytic subset of X. We set for short $\mathcal{C}^{\infty,\mathrm{w}}_{X|S}$ instead of $\mathcal{C}^{\infty,\mathrm{w}}_{X|\mathbb{C}_S}$.*

Let $\chi = \{M_1,\dots,M_\ell\}$ be a family of closed analytic submanifolds of X, set $\bigcup_k M = \bigcup_{k=1}^\ell M_k$ and consider $\widetilde{X}$. Set $F = \mathbb{C}_{X\setminus\bigcup_k M}$, $G = \mathbb{C}_X$, $H = \mathbb{C}_{\bigcup_k M}$ in (4.21). The exact sequence

$$0 \to \mathcal{C}^{\infty,\mathrm{w}}_{X|X\setminus\bigcup_k M} \to \mathcal{C}^{\infty,\mathrm{w}}_X \to \mathcal{C}^{\infty,\mathrm{w}}_{X|\bigcup_k M} \to 0$$

induces an exact sequence

$$0 \to \rho^{-1}\nu^{sa}_\chi \mathcal{C}^{\infty,\mathrm{w}}_{X|X\setminus\bigcup_k M} \to \rho^{-1}\nu^{sa}_\chi \mathcal{C}^{\infty,\mathrm{w}}_X \to \rho^{-1}\nu^{sa}_\chi \mathcal{C}^{\infty,\mathrm{w}}_{X|\bigcup_k M} \to 0, \tag{4.22}$$

where the surjective arrow is the map which associates to a function its asymptotic expansion. In fact, let V be a l.c.t. conic subanalytic subset of the zero section of $\widetilde{X}$ and $U \in \mathrm{Op}(X_{sa})$ such that $C_\chi(X\setminus U)\cap V = \emptyset$, then we can find a l.c.t. $U' \subset U$ satisfying the same property.

Let X be a complex manifold ($X = \mathbb{C}^n$ in the local model) and let $X_\mathbb{R}$ denote the underlying real analytic manifold of X. Let $\chi = \{Z_1,\dots,Z_\ell\}$ be a family of complex submanifolds of X and set $\bigcup_k Z = \bigcup_{k=1}^\ell Z_k$. Let $F \in D^b_{\mathbb{R}\text{-c}}(\mathbb{C}_X)$. We denote by $\mathcal{O}^{\mathrm{w}}_{X|F}$ the sheaf defined as follows:

$$\mathcal{O}^{\mathrm{w}}_{X|F} := R\mathcal{H}om_{\rho_!\mathcal{D}_{\overline{X}}}(\rho_!\mathcal{O}_{\overline{X}}, \mathcal{C}^{\infty,\mathrm{w}}_{X_\mathbb{R}|F}).$$

Let $0 \to F \to G \to H \to 0$ be an exact sequence in $\mathrm{Mod}_{\mathbb{R}\text{-c}}(\mathbb{C}_X)$. Then the exact sequence (4.21) gives rise to the distinguished triangle

$$\mathcal{O}^{\mathrm{w}}_{X|F} \to \mathcal{O}^{\mathrm{w}}_{X|G} \to \mathcal{O}^{\mathrm{w}}_{X|H} \overset{+}{\to} . \tag{4.23}$$

Setting $F = \mathbb{C}_{X\setminus\bigcup_k Z}$, $G = \mathbb{C}_X$, $H = \mathbb{C}_{\bigcup_k Z}$ in (4.23) and applying the functor of specialization, we have the distinguished triangle

$$\rho^{-1}\nu^{sa}_\chi \mathcal{O}^{\mathrm{w}}_{X|X\setminus\bigcup_k Z} \to \rho^{-1}\nu^{sa}_\chi \mathcal{O}^{\mathrm{w}}_X \to \rho^{-1}\nu^{sa}_\chi \mathcal{O}^{\mathrm{w}}_{X|\bigcup_k Z} \overset{+}{\to} . \tag{4.24}$$

Let us consider the multi-specialization of Whitney holomorphic functions. Let W be an $(\mathbb{R}^+)^\ell$-conic subanalytic open subset in S_χ. Remember that $(\mathbb{R}^+)^\ell$-conic open subset V' in S_χ is said to be compactly generated in W if there exists a compact subset K in W with $V' \subset (\mathbb{R}^+)^\ell K$.

We have

$$H^0(W; \rho^{-1}\nu^{sa}_\chi \mathcal{O}^{\mathrm{w}}_X) = \varprojlim_{V'} \varinjlim_{U'} H^0(U'; \mathcal{O}^{\mathrm{w}}_X),$$

where V' ranges through the family of $(\mathbb{R}^+)^\ell$-conic subanalytic open subsets in S_χ which are compactly generated in W, U' ranges through the family of $\mathrm{Op}(X_{sa})$ such that $C_\chi(X \setminus U') \cap V' = \emptyset$. One can check easily that, given $p = (0; \zeta) \in S_\chi$, then

$$\begin{aligned}(\rho^{-1}H^0\nu^{sa}_\chi \mathcal{O}^{\mathrm{w}}_X)_p &\simeq \varinjlim_{U} H^0(U; \nu^{sa}_\chi \mathcal{O}^{\mathrm{w}}_X) \\ &\simeq \varinjlim_{V,\epsilon} H^0(S(V,\epsilon); \mathcal{O}^{\mathrm{w}}_X) \\ &\simeq \varinjlim_{V,\epsilon} H^0(X; \mathbb{C}_{\overline{S(V,\epsilon)}} \overset{\mathrm{w}}{\otimes} \mathcal{O}_X),\end{aligned}$$

where U ranges through the $(\mathbb{R}^+)^\ell$-conic neighborhoods of p, V through the family $V(\zeta)$ and $\epsilon > 0$. The last isomorphism follows from Corollary 4.9. Hence we can construct functorially the sheaf $\rho^{-1}H^0\nu^{sa}_\chi \mathcal{O}^{\mathrm{w}}_X$ and its stalk can be described by multi-asymptotics, which is introduced in [9], [10] and [11].

As a matter of fact, let U be an open l.c.t. subanalytic subset in X. Then $H^0(U; \mathcal{O}^{\mathrm{w}}_X) \simeq H^0(X; \mathbb{C}_{\overline{U}} \overset{\mathrm{w}}{\otimes} \mathcal{O}_X)$ consists of $\mathcal{C}^\infty$-Whitney jets on $\overline{U}$ that satisfy the Cauchy-Riemann system. Therefore, if a proper cone $S := S(V, \epsilon)$ is subanalytic, in view of Theorem 7.13 [9] for χ with the condition H2 and Theorem 3.37 [10] for a general χ, the set of holomorphic functions on S that are multi-asymptotically developable along χ is equal to

$$\varprojlim_{S'} H^0(X; \mathbb{C}_{\overline{S'}} \overset{\mathrm{w}}{\otimes} \mathcal{O}_X),$$

where S' runs through the family of open subanalytic proper cones $S(V', \epsilon')$ properly contained in S.

Similarly we have

$$(\rho^{-1}H^0\nu^{sa}_\chi \mathcal{O}^{\mathrm{w}}_{X|X\setminus\bigcup_k Z})_p \simeq \varinjlim_{V,\epsilon} H^0(X; \mathbb{C}_{\overline{S(V,\epsilon)}\setminus\bigcup_k Z} \overset{\mathrm{w}}{\otimes} \mathcal{O}_X),$$

where V through the family $V(\zeta)$ and $\epsilon > 0$. Hence we can construct functorially the sheaf $\rho^{-1}H^0\nu^{sa}_\chi \mathcal{O}^{\mathrm{w}}_{X|X\setminus\bigcup_k Z}$ and its stalk gives "flat multi-asymptotics" introduced in [9], [10] and [11]. In fact, it follows from Proposition 7.14 [9] for χ with the condition H2 and Theorem 3.49 [10] for a general χ that the set of holomorphic functions on a proper subanalytic cone $S = S(V, \epsilon)$ that are multi-asymptotically developable along χ and flat on S is equal to

$$\varprojlim_{S'} H^0(X; \mathbb{C}_{\overline{S'} \setminus \bigcup_k Z} \overset{\mathrm{w}}{\otimes} \mathcal{O}_X),$$

where S' runs through the family of open subanalytic proper cones $S(V', \epsilon')$ properly contained in S.

Finally, one can check

$$(\rho^{-1} H^0 \nu_\chi^{sa} \mathcal{O}^{\mathrm{w}}_{X | \bigcup_k Z})_p \simeq \varinjlim_{V, \epsilon} H^0(X; \mathbb{C}_{\overline{S(V,\epsilon)} \cap \bigcup_k Z} \overset{\mathrm{w}}{\otimes} \mathcal{O}_X),$$

where V through the family $V(\zeta)$ and $\epsilon > 0$. Hence we can construct functorially the sheaf $\rho^{-1} H^0 \nu_\chi^{sa} \mathcal{O}^{\mathrm{w}}_{X|\bigcup_k Z}$. We also remark that its stalk can be described by the language of multi-asymptotics, that is, "consistent families of coefficients" introduced in [9], [10] and [11]. In fact, by Proposition 7.18 [9] for χ with the condition H2 and Theorem 3.45 [10] for a general χ, we see that the set of total families of coefficients of multi-asymtotic expansion along χ on a proper subanalytic cone $S = S(V, \epsilon)$ which are consistent is equal to

$$\varprojlim_{S'} H^0(X; \mathbb{C}_{\overline{S'} \cap \bigcup_k Z} \overset{\mathrm{w}}{\otimes} \mathcal{O}_X),$$

where S' runs through the family of open subanalytic proper cones $S(V', \epsilon')$ properly contained in S.

5 Vanishing theorems

In this section, we establish several vanishing theorems related to multi-specializations of holomorphic functions with growth conditions.

5.1 Vanishing theorems for Whitney holomorphic functions and Borel-Ritt exact sequence

In this subsection, we show several vanishing theorems related to multi-specializations of the sheaves of Whitney holomorphic functions. As an application we obtain a Borel-Ritt exact sequence for multi-asymptotics. We continue to consider the problems in the complex domain with the same geometrical situation introduced in Subsection 4.1. We also continue to use the same notations as those in Subsection 4.1.

Let $X = \mathbb{C}^n$ with coordinates $z = (z_1, \dots, z_n)$ and

$$p = (0, \zeta) = (0, \zeta^{(1)}, \dots, \zeta^{(m)}) \in S_X.$$

We first recall the definitions of J_k and I_j: These are subsets defined by

$$J_k = \{j \in \{1, 2, \cdots, \ell\};\ \nu_j^\lambda(\varphi_k) > 0\} \qquad (k = 1, 2, \cdots, m)$$

and

$$I_j = \{k \in \{1, 2, \cdots, m\} \mid \nu_j^\lambda(\varphi_k) > 0\} \quad (j = 1, 2, \cdots, \ell).$$

We also define the subset Z_j in $(\mathbb{R}_{\geq 0})^m$ by

$$\{\tau \in (\mathbb{R}_{\geq 0})^m\,;\ \tau_k = 0 \quad (k \in I_j)\} \qquad (j = 1, 2, \cdots, \ell).$$

Hereafter we use the same symbol Z_j for the subset of $(\mathbb{R}_{\geq 0})^m$ defined above and the one of X already introduced in the previous sections if there is no confusion.

Let H be a finite subset of $\mathcal{H}_\tau \times \mathbb{R}_{\geq 0}$ such that the semigroups $[H]$ and $\mathcal{G}$ are equivalent. Furthermore, we assume the condition $Q(H) = H$ (it is easily satisfied by replacing H with $Q(H)$). Note that, as $[H]$ and $\mathcal{G}$ are equivalent, for $k \in J_Z(\zeta)$ and for any $(f, v) \in H$, we have $\nu_k(f) \geq 0$ and $\nu_k(f) = 0$ if $v \neq 0$.

Define

$$I_D(H) = \{i \in \{1, 2, \cdots, m\}\,;\ \nu_i(f) < 0 \text{ for some } (f, v) \in H\},$$

that is, $I_D(H)$ is the set of indices k such that the variable τ_k appears in the denominator of some f $((f, v) \in H)$. Note that $I_D(H)$ is independent of the choices of H, that is, if $[H]$ and $[L]$ are equivalent, then $I_D(H) = I_D(L)$ holds.

Let $\{\epsilon_{(f,v)}\}_{(f,v) \in H}$ be a set of real numbers with $\epsilon_{(f,v)} > 0$. We define

$$H_\epsilon = \{(f, v + \epsilon_{(f,v)});\ (f, v) \in H\}.$$

For $k \in \{1, 2, \cdots, m\}$, we consider the operation $\widehat{\mathcal{L}}_k$ on $\mathcal{H}_\tau \times \mathbb{R}_{\geq 0}$ consisting of only F1 and F3 (i.e., without F2), which is similar to the operation $\mathcal{L}_j^\lambda$. Further, we define the operation $\widehat{\mathcal{L}}_k^*$ by

$$\widehat{\mathcal{L}}_k^*(L) = \widehat{\mathcal{L}}_k(L) \bigcup L$$

for a subset $L \subset \mathcal{H}_\tau \times \mathbb{R}_{\geq 0}$.

Let $T' \subset \{1, 2, \cdots, m\}$ and $L \subset \mathcal{H}_\tau \times \mathbb{R}_{\geq 0}$. We recursively define $\varpi_{T'}(L) \subset \mathcal{H}_\tau \times \mathbb{R}_{\geq 0}$ in the following way: For $T' = \{1, 2, \cdots, m\}$, we set

$$\varpi_{T'}(L) = L.$$

Then we recursively define

$$\varpi_{T'}(L) = \bigcup_{k \in \{1,2,\cdots,m\} \setminus T'} \widehat{\mathcal{L}}_k^*(\varpi_{T' \cup \{k\}}(L)).$$

We set

$$\varpi(L) = \varpi_{\emptyset}(L).$$

The most important property of $\varpi(L)$ is as follows: For $k \in \{1, 2, \ldots, m\}$, we have

$$(f, v),\ (g, w) \in \varpi(L) \text{ with } \nu_k(f) > 0,\ \nu_k(g) < 0 \implies (f^a g^b, v^a w^b) \in \varpi(L),$$

where a, b are natural numbers prime to each other and they satisfy $a|\nu_k(f)| = b|\nu_k(g)|$, that is, $\varpi(L)$ is stable under the operation F3, i.e.,

$$\widehat{\mathcal{L}}_k^*(\varpi(L)) \subset \varpi(L).$$

Furthermore, for a subset $T \subset \{1, 2, \cdots, m\}$, we define

$$\varpi(L)_T = \{(f, v) \in \varpi(L)\,;\, \nu_k(f) = 0\ (k \notin T)\}.$$

Note that, by the above observation, we have

$$\widehat{\mathcal{L}}_k(\varpi(L)_T) \subset \varpi(L)_{T \setminus \{k\}} \tag{5.1}$$

for $k \in T$.

Set

$$S(L) = \{\tau \in (\mathbb{R}_{>0})^m;\ f(\tau) < v \text{ for any } (f, v) \in L\}.$$

Note that if $L = Q(L)$ holds, then for any $(f, v) \in L$ with $v \neq 0$, $(1/f, 1/v)$ also belongs to L. Hence inequalities of a kind $v - \epsilon_{(f,v),-} < f(\tau)$ are also taken into account in the above expression of $S(L)$. We also set

$$\widetilde{S}(L) = \{\tau \in (\mathbb{R}_{\geq 0})^m;\ f_N(\tau) \leq v f_D(\tau) \text{ for any } (f = f_N/f_D, v) \in L\},$$

where f_N and f_D are monomials whose exponents are non-negative rational numbers and they have no common variables. In what follows, we write $f = f_N/f_D$ in this sense. One should aware that $S(L)$ is a subset of $(\mathbb{R}_{>0})^m$, but $\widetilde{S}(L)$ is the one of $(\mathbb{R}_{\geq 0})^m$.

Proposition 5.1. *Under the above situation, we have* $\widetilde{S}(\varpi(H_\epsilon)) = \overline{S(H_\epsilon)}$.

Before the proof of the proposition, we need some preparations: Let T be a subset in $\{1, 2, \cdots, m\}$. Set $\tau' = (\tau_k)_{k \in T}$ and

$$\mathbb{R}^T_{\geq 0} = \{(\tau') \in \mathbb{R}^{\#T} \,|\, \tau_k \geq 0\ (k \in T)\},$$
$$\mathbb{R}^T_{>0} = \{(\tau') \in \mathbb{R}^{\#T} \,|\, \tau_k > 0\ (k \in T)\}.$$

We denote by $\pi_T : \mathbb{R}^m \to \mathbb{R}^{\#T}$ the canonical projection $\tau \mapsto \tau'$. In the same way, for any subsets $T' \subset T \subset \{1, 2, \cdots, m\}$, we can define the canonical projection $\pi_{T',T} : \mathbb{R}^{\#T} \to \mathbb{R}^{\#T'}$ by $\tau' \mapsto \tau'' = (\tau_j)_{j \in T'}$. Let $L \subset \mathcal{H}_{\tau'} \times \mathbb{R}_{\geq 0}$. We set

$$S_T(L) = \{(\tau') \in \mathbb{R}^T_{>0}; f(\tau') < v \quad ((f, v) \in L)\}$$

and

$$\widetilde{S}_T(L) = \left\{\tau' \in \mathbb{R}^T_{\geq 0};\ f_N(\tau') \leq v f_D(\tau') \quad ((f = f_N/f_D, v) \in L)\right\}.$$

Lemma 5.2. *1. We have, for any $k \in I_D(H) = I_D(\varpi(H_\epsilon))$,*

$$\widetilde{S}(\varpi(H_\epsilon)) \bigcap \{\tau_k = 0\} \subset \bigcup_{j \in J_k} Z_j.$$

2. If $(1, v) \in [H_\epsilon]$, then we have $v > 1$.

3. For any subsets $T \subset T' \subset \{1, 2, \cdots, m\}$, we have

$$S_T(\varpi(H_\epsilon)_T) = \pi_{T,T'}(S_{T'}(\varpi(H_\epsilon)_{T'})) = \pi_T(S(H_\epsilon)).$$

Proof. First we will show the claim 1. of the lemma: We consider $\mathbb{Z}_2 = \{0, 1\} \subset \mathbb{R}_{\geq 0}$ with the order $0 < 1$ and define $\sigma : \mathbb{R}_{\geq 0} \to \mathbb{Z}_2$ by $\sigma(a) = 1$ if $a > 0$ and $\sigma(0) = 0$. Since $\sigma(ab) = \sigma(a)\sigma(b)$ and

$$a \leq b \Longrightarrow \sigma(a) \leq \sigma(b)$$

hold, we have

$$(\tau_1, \cdots, \tau_m) \in \widetilde{S}(\varpi(H_\epsilon)) \Longrightarrow (\sigma(\tau_1), \cdots, \sigma(\tau_m)) \in S(\mathbb{Z}_2;\ \varpi(H_\epsilon)) = S(\mathbb{Z}_2;\ \varpi(H)),$$

where we set, for a subset $L \subset \mathcal{H}_\tau \times \mathbb{R}_{\geq 0}$,

$$S(\mathbb{Z}_2;\ L) = \{\tau \in (\mathbb{Z}_2)^m;\ f_N(\tau) \leq f_D(\tau) \quad ((f = f_N/f_D, v) \in L)\}.$$

Furthermore $\sigma(a) = 0 \iff a = 0$ holds. Hence it suffices to show, for $k \in I_D(H)$,

$$S(\mathbb{Z}_2;\ \varpi(H)) \bigcap \{\tau_k = 0\} \subset \bigcup_{j \in J_k} Z_j.$$

We remark that, if $[L]$ and $[H]$ are equivalent, then we have

$$I_D(H) = I_D(L) \quad \text{and} \quad S(\mathbb{Z}_2;\ \varpi(H)) = S(\mathbb{Z}_2;\ \varpi(L)).$$

As a matter of fact, for any $(f, v) \in \varpi(H)$, we can find an expression

$$(f, v) = \prod_{i=1,\cdots,r} (g_i, w_i)^{\alpha_i},$$

where

1. $\alpha_i \in \mathbb{Q}_{\geq 0}$ and $(g_i, w_i) \in \varpi(L)$.
2. there exists no pair i, j such that $\nu_k(g_i) < 0$ and $\nu_k(g_j) > 0$ for some k. In fact, starting from $k = 1$ to m, if such a pair exists for the variable τ_k, we can successively eliminate the pair by the operation F3.

Hence it follows from the above condition 2. that we have

$$g_{i,N}(\tau) \leq g_{i,D}(\tau) \ (i = 1, 2, \cdots, r) \implies f_N(\tau) \leq f_D(\tau),$$

where $f = f_N/f_D$ and $g_i = g_{i,N}/g_{i,D}$ $(i = 1, 2, \cdots, r)$. Note that, without the condition 2., the above implication generally does not hold. As a conclusion, we get

$$S(\mathbb{Z}_2;\, \varpi(H)) \supset S(\mathbb{Z}_2;\, \varpi(L)).$$

The converse inclusion also holds by exchanging the roles of H and L, and thus, we have obtained $S(\mathbb{Z}_2;\, \varpi(H)) = S(\mathbb{Z}_2;\, \varpi(L))$.

By the above remark, we may take the following special finite subset as H: Recall the definition of $\widehat{G} \subset \mathcal{H}_{\tau,\lambda} \times \mathbb{R}_{\geq 0}$, that is,

$$\widehat{G} = Q\left(\left\{\left(\frac{\tau_1}{\varphi_1(\lambda)}, |\zeta^{(1)}|\right), \cdots, \left(\frac{\tau_m}{\varphi_m(\lambda)}, |\zeta^{(m)}|\right), (\lambda_1, 0), \cdots, (\lambda_\ell, 0)\right\}\right).$$

Now we apply the operations $\mathcal{L}_k^\lambda$ $(k = 1, \cdots, \ell)$ to $\widehat{G}$ successively, we get

$$(\mathcal{L}_1^\lambda \circ \cdots \circ \mathcal{L}_\ell^\lambda)(\widehat{G}) \subset \mathcal{H}_\tau \times \mathbb{R}_{\geq 0}$$

which is chosen as H hereafter. Note that $[H]$ and $\mathcal{G}$ are equivalent.

Furthermore, since $S(\mathbb{Z}_2;\, \varpi(H)) \subset S(\mathbb{Z}_2;\, H)$ holds, it suffices to show, for $k \in I_D(H)$,

$$S(\mathbb{Z}_2;\, H) \bigcap \{\tau_k = 0\} \subset \bigcup_{j \in J_k} Z_j.$$

Let us show the above inclusion: Set $\widehat{H}_\ell = \widehat{G}$ and

$$\widehat{H}_j := (\mathcal{L}_{j+1}^\lambda \circ \cdots \circ \mathcal{L}_\ell^\lambda)(\widehat{G}) \qquad (0 \leq j \leq \ell - 1).$$

Note that $H = \widehat{H}_0$. We define $I_D(L)$ for $L \subset \mathcal{H}_{\tau,\lambda} \times \mathbb{R}_{\geq 0}$ in the same way as that for a subset in $\mathcal{H}_\tau \times \mathbb{R}_{\geq 0}$, i.e.,

$$I_D(L) = \{i \in \{1, 2, \cdots, m\}\,;\, \nu_i(f) < 0 \text{ for some } (f, v) \in L\}.$$

We also define the set $S_{m,j}(\mathbb{Z}_2;\, \widehat{H}_j)$ by

$$\left\{(\tau, \lambda') \in (\mathbb{Z}_2)^m \times (\mathbb{Z}_2)^j;\, f_N(\tau, \lambda') \leq f_D(\tau, \lambda') \quad ((f = f_N/f_D, v) \in \widehat{H}_j)\right\},$$

where $\lambda' = (\lambda_1, \cdots, \lambda_j)$ and

$$\widehat{Z}_j = \{\tau_k = 0 \ (k \in I_j)\} \cap \{\lambda_j = 0\} \qquad (j = 1, 2, \cdots, \ell).$$

Then we will show the following claim, for any $0 \le p \le \ell$ and $k \in I_D(\widehat{H}_p)$

$$S_{m,p}(\mathbb{Z}_2;\, \widehat{H}_p) \bigcap \{\tau_k = 0\} \subset \left(\bigcup_{j \in J_k,\, j \le p} \widehat{Z}_j \right) \bigcup \left(\bigcup_{j \in J_k,\, p+1 \le j} Z_j \right).$$

If $p = \ell$, then the claim is, for $k \in I_D(\widehat{G})$,

$$S_{m,\ell}(\mathbb{Z}_2;\, \widehat{G}) \bigcap \{\tau_k = 0\} \subset \bigcup_{j \in J_k} \widehat{Z}_j,$$

which clearly holds by the definition of $\widehat{G}$.

Now let us show the claim for $p = \ell - 1$. Take $(\tau, \lambda') \in S_{m,\ell-1}(\mathbb{Z}_2;\, \widehat{H}_{\ell-1})$. We set $\lambda_\ell = 1 \in \mathbb{Z}_2$ if there exists an $(f = f_N/f_D, v) \in \widehat{H}_\ell$ with $\nu_\ell^\lambda(f) < 0$ such that

$$f_N(\tau, \lambda') > 0 \qquad (\text{i.e., } f_N(\tau, \lambda') = 1).$$

Otherwise, we set $\lambda_\ell = 0 \in \mathbb{Z}_2$. Then, by inequalities associated with terms generated by the operation F3 and the fact that $(\lambda_\ell, 0) \in \widehat{H}_\ell$, we can see

$$(\tau, \lambda) = (\tau, \lambda', \lambda_\ell) \in S_{m,\ell}(\mathbb{Z}_2;\, \widehat{H}_\ell), \tag{5.2}$$

which is shown in the following way:

1. When $\lambda_\ell = 1$, there exists an $(f = f_N/f_D, v) \in \widehat{H}_\ell$ with $\nu_\ell^\lambda(f) < 0$ and $f_N(\tau, \lambda') = 1$. Here, since $(f, v)(\lambda_\ell, 0)^{\nu_\ell^\lambda(f)} \in \widehat{H}_{\ell-1}$ and $\lambda_\ell = 1$ hold, we also have $f_D(\tau, \lambda) = 1$.
   Let $(g = g_N/g_D, w) \in \widehat{H}_\ell$. If $\nu_\ell^\lambda(g) > 0$, then we have $(g, w)^a (f, v)^b \in \widehat{H}_{\ell-1}$ for some $a, b \in \mathbb{N}$. Therefore, $f_N(\tau, \lambda') = f_D(\tau, \lambda) = 1$ imply $g_N(\tau, \lambda) \le g_D(\tau, \lambda)$. If $\nu_\ell^\lambda(g) < 0$, then $(g, w)(\lambda_\ell, 0)^{\nu_\ell^\lambda(g)} \in \widehat{H}_{\ell-1}$ and $\lambda_\ell = 1$ imply $g_N(\tau, \lambda) \le g_D(\tau, \lambda)$. These conclude (5.2).
2. For the case $\lambda_\ell = 0$, (5.2) immediately follows from the condition of setting λ_ℓ to 0.

Hence, it follows from the claim for $p = \ell$ and the fact $I_D(\widehat{H}_{\ell-1}) \subset I_D(\widehat{H}_\ell)$ that we have, for $k \in I_D(\widehat{H}_{\ell-1})$,

$$\begin{aligned} S_{m,\ell-1}(\mathbb{Z}_2;\, \widehat{H}_{\ell-1}) \bigcap \{\tau_k = 0\} &\subset \left(\bigcup_{j \in J_k,\, j \le \ell-1} \widehat{Z}_j \right) \bigcup \begin{cases} Z_\ell & (\ell \in J_k), \\ \emptyset & (\ell \notin J_k) \end{cases} \\ &= \left(\bigcup_{j \in J_k,\, j \le \ell-1} \widehat{Z}_j \right) \bigcup \left(\bigcup_{j \in J_k,\, \ell \le j} Z_j \right). \end{aligned}$$

By repetition of the same arguments, we can get the claim for $p = 0$. This completes the proof of the claim 1. of the lemma.

Let us show the claim 2. of the lemma: For $(f, v) \in [H]$, we have the unique expression with $\alpha_i \in \mathbb{Q}$ and $\beta_j \in \mathbb{Q}_{\geq 0}$,

$$(f, v) = \left(\frac{\tau_1}{\varphi_1(\lambda)}, |\zeta^{(1)}|\right)^{\alpha_1} \cdots \left(\frac{\tau_m}{\varphi_m(\lambda)}, |\zeta^{(m)}|\right)^{\alpha_m} (\lambda_1, 0)^{\beta_1} \cdots (\lambda_\ell, 0)^{\beta_\ell}.$$

Therefore $f = 1$ implies $\alpha_i = \beta_j = 0$, and thus, we get $v = 1$.

Then, for $(f, v) \in [H_\epsilon]$, we have

$$(f, v) = (f_1, v_1 + \epsilon_{(f_1, v_1)})^{\alpha_i} \cdots (f_r, v_r + \epsilon_{(f_r, v_r)})^{\alpha_r},$$

where $(f_i, v_i) \in H$ and $\alpha_i \in \mathbb{Z}_{\geq 0}$. It follows from the above observation that, if $f = 1$, then $v_1^{\alpha_1} \cdots v_r^{\alpha_r} = 1$. Hence we conclude $v > 1$.

Finally we will show the claim 3. of the lemma: It is enough to show

$$S_T(\varpi(H_\epsilon)_T) = \pi_{T,T'}(S_{T'}(\varpi(H_\epsilon)_{T'}))$$

because the other inequalities come from this. As $\varpi(H_\epsilon)_T \subset \varpi(H_\epsilon)_{T'}$ holds, we have

$$\pi_{T,T'}(S_{T'}(\varpi(H_\epsilon)_{T'})) \subset S_T(\varpi(H_\epsilon)_T).$$

Hence we will show

$$S_T(\varpi(H_\epsilon)_T) \subset \pi_{T,T'}(S_{T'}(\varpi(H_\epsilon)_{T'})). \tag{5.3}$$

Apparently the problem can be reduced to the case $T = \{1, 2, \cdots, d-1\}$ and $T' = \{1, 2, \cdots, d\}$ $(d \geq 1)$. In this case, we have

$$S_{T'}(\varpi(H_\epsilon)_{T'}) = \{\tau = (\tau', \tau_d) \in \mathbb{R}^d_{>0};\, \tau' \in S_T(\varpi(H_\epsilon)_T),\ \varphi_L(\tau') < \tau_d < \varphi_H(\tau')\}.$$

Here $\tau' = (\tau_1, \cdots, \tau_{d-1})$ and we set

$$\varphi_L(\tau') = \max_{(g,w)} (g_N/(wg'_D))^{1/|\nu_d(g)|},$$

where $(g = \tau_d^{\nu_d(g)}(g_N/g'_D), w)$ runs through

$$(\varpi(H_\epsilon)_{T'})_- = \{(g, w) \in \varpi(H_\epsilon)_{T'};\, \nu_d(g) < 0\},$$

and we also set

$$\varphi_H(\tau') = \min_{(f,v)} (vf_D/f'_N)^{1/|\nu_d(f)|},$$

where $(f = \tau_d^{\nu_d(f)}(f'_N/f_D), v)$ ranges through

$$(\varpi(H_\epsilon)_{T'})_+ = \{(f, v) \in \varpi(H_\epsilon)_{T'};\, \nu_d(f) > 0\}.$$

Let (f, v) and (g, w) be in $\varpi(H_\epsilon)_{T'}$ with $\nu_d(f) > 0$ and $\nu_d(g) < 0$. Then it follows from (5.1) that we have

$$(f^a g^b, v^a w^b) \in \varpi(H_\epsilon)_T,$$

where natural numbers a and b are prime to each other with $a|\nu_d(f)| = b|\nu_d(g)|$. Hence we get

$$(f^a \cdot g^b)(\tau') < v^a w^b \qquad (\tau' \in S_T(\varpi(H_\epsilon)_T)),$$

which guarantees

$$\varphi_L(\tau') < \varphi_H(\tau') \qquad (\tau' \in S_T(\varpi(H_\epsilon)_T)).$$

This shows (5.3), which completes the proof. □

Now we will show the proposition.

Proof. Since we have $S(H_\epsilon) = S(\varpi(H_\epsilon))$, it suffices to show

$$\overline{S(\varpi(H_\epsilon))} = \widetilde{S}(\varpi(H_\epsilon)).$$

Set

$$\widetilde{H} = \varpi(H_\epsilon), \qquad \widetilde{H}_T = \varpi(H_\epsilon)_T.$$

Clearly we have $\overline{S(\widetilde{H})} \subset \widetilde{S}(\widetilde{H})$. Hence it suffices to show $\overline{S(\widetilde{H})} \supset \widetilde{S}(\widetilde{H})$. Let $\tau^* = (\tau_1^*, \cdots, \tau_m^*) \in \widetilde{S}(\widetilde{H})$ and set

$$T = \{1, 2, \cdots, m\} \setminus (J_Z(\tau^*) \cap I_D(H)),$$

where $J_Z(\tau^*)$ denotes the subset $\{k \in \{1, 2, \cdots, m\};\ \tau_k^* = 0\}$. Note that

$$(J_Z(\tau^*) \cap T) \cap I_D(H) = \emptyset$$

holds. We may assume

$$T = \{1, 2, \cdots, m'\} \qquad (m' \leq m).$$

For $d = 0, 1, \cdots, m'$, let T_d denote the subset $\{1, 2, \cdots, d\}$. We first show the claim

$$\overline{S_{T_d}(\widetilde{H}_{T_d})} = \widetilde{S}_{T_d}(\widetilde{H}_{T_d}) \quad \text{near } \pi_{T_d}(\tau^*)$$

by the induction with respect to d. Apparently we have

$$\overline{S_{T_d}(\widetilde{H}_{T_d})} \subset \widetilde{S}_{T_d}(\widetilde{H}_{T_d}).$$

Hence it suffices to show the converse inclusion near $\pi_{T_d}(\tau^*)$.

Let $d=0$, i.e., $T_0=\emptyset$. It follows from the claim 2. of Lemma 5.2 that we have

$$\overline{S_{T_0}(\widetilde{H}_{T_0})} = S_{T_0}(\widetilde{H}_{T_0}) = \widetilde{S}_{T_0}(\widetilde{H}_{T_0}) = \{0\}.$$

Therefore, the claim holds in this case.

We assume that the claim holds for $d-1$ ($d \geq 1$). Then, using notations in the proof of the claim 3. of Lemma 5.2, we have

$$S_{T_d}(\widetilde{H}_{T_d}) = \{(\tau', \tau_d) \in (\mathbb{R}_{>0})^d\,;\, \tau' \in S_{T_{d-1}}(\widetilde{H}_{T_{d-1}}),\ \varphi_L(\tau') < \tau_d < \varphi_H(\tau')\}.$$

Since the denominator of f for $(f, v) \in \widetilde{H}_{T_d}$ never becomes zero at $\pi_{T_d}(\tau^*)$ by the assumption $(J_Z(\tau^*) \cap T) \cap I_D(H) = \emptyset$, $\varphi_L(\tau')$ becomes a continuous function on $(\mathbb{R}^n_{\geq 0})^{d-1}$ near $\tau' = \pi_{T_{d-1}}(\tau^*)$. In the same way, $\varphi_H(\tau')$ also becomes a continuous function from $(\mathbb{R}_{\geq 0})^{d-1}$ to $[0, +\infty]$ near $\tau' = \pi_{T_{d-1}}(\tau^*)$. Note that we use the convention

$$a(+\infty) = a/0 = +\infty \ (a > 0) \quad \text{and} \quad a/(+\infty) = 0 \ (a \geq 0),$$

under which $\varphi_H(\tau')$ is allowed to take the value $+\infty$. Therefore, by the continuity of φ_L and φ_H near $\pi_{T_{d-1}}(\tau^*)$ and the induction hypothesis, we have near $\pi_{T_d}(\tau^*)$

$$\begin{aligned}\overline{S_{T_d}(\widetilde{H}_{T_d})} &= \{(\tau', \tau_d) \in (\mathbb{R}_{\geq 0})^d\,;\, \tau' \in \overline{S_{T_{d-1}}(\widetilde{H}_{T_{d-1}})},\ \varphi_L(\tau') \leq \tau_d \leq \varphi_H(\tau')\} \\ &= \{(\tau', \tau_d) \in (\mathbb{R}_{\geq 0})^d\,;\, \tau' \in \widetilde{S}_{T_{d-1}}(\widetilde{H}_{T_{d-1}}),\ \varphi_L(\tau') \leq \tau_d \leq \varphi_H(\tau')\}.\end{aligned}$$

Further, we have the following fact: For $(f, v) \in (\widetilde{H}_{T_d})_+$ with $f = \tau_d^{\nu_d(f)}(f'_N/f_D)$, since $f_D(\pi_{T_d}(\tau^*)) \neq 0$ holds by the assumption, we have the equivalence

$$\tau_d \leq (v f_D(\tau', \tau_d)/f'_N(\tau'))^{1/|\nu_d(f)|} \iff \tau_d^{\nu_d(f)} f'_N(\tau') \leq v f_D(\tau', \tau_d)$$

near $\pi_{T_d}(\tau^*)$. We have also the similar equivalence for $(g, w) \in (\widetilde{H}_{T_d})_-$.

By these observations, we have obtained

$$\begin{aligned}\overline{S_{T_d}(\widetilde{H}_{T_d})} &= \{(\tau', \tau_d) \in (\mathbb{R}_{\geq 0})^d\,;\, \tau' \in \widetilde{S}_{T_{d-1}}(\widetilde{H}_{T_{d-1}}),\ \varphi_L(\tau') \leq \tau_d \leq \varphi_H(\tau')\} \\ &\supset \{(\tau', \tau_d) \in (\mathbb{R}_{\geq 0})^d;\ f_N(\tau', \tau_d) \leq v f_D(\tau', \tau_d) \quad ((f = f_N/f_D, v) \in \widetilde{H}_{T_d})\} \\ &= \widetilde{S}_{T_d}(\widetilde{H}_{T_d}) \quad \text{near } \pi_{T_d}(\tau^*),\end{aligned}$$

which shows the claim. That is, we have obtained

$$\overline{S_T(\widetilde{H}_T)} = \widetilde{S}_T(\widetilde{H}_T) \quad \text{near } \pi_T(\tau^*). \tag{5.4}$$

Let us denote by $\mu_j(\lambda; \tau)$ the action on $(\mathbb{R}_{\geq 0})^m$ defined by

$$\mu_j(\tau;\, \lambda) = (\varphi_1(\tilde{\lambda})\tau_1, \cdots, \varphi_m(\tilde{\lambda})\tau_m),$$

where $\tilde{\lambda} = (1, \cdots, 1, \overset{j\text{-th}}{\lambda}, 1, \cdots, 1)$.

Let $k \in J_Z(\tau^*) \cap I_D(H)$. Then it follows from the claim 1. of Lemma 5.2 that we can find $j \in J_k$ such that $\tau^* \in Z_j$. Note that

$$j \in J_k \iff k \in I_j, \qquad \tau^* \in Z_j \iff I_j \subset J_Z(\tau^*)$$

hold. Hence, applying the above argument to each index in $J_Z(\tau^*) \cap I_D(H)$, we can find $j_1, \cdots, j_d \in \{1, 2, \cdots, \ell\}$ such that

$$\begin{aligned} &\left(\bigcup_{k=1}^{d} I_{j_k}\right) \bigcap I_D(H) = J_Z(\tau^*) \cap I_D(H), \\ &\left(\bigcup_{k=1}^{d} I_{j_k}\right) \bigcap (\{1, 2, \cdots, m\} \setminus I_D(H)) \subset J_Z(\tau^*) \bigcap (\{1, 2, \cdots, m\} \setminus I_D(H)). \end{aligned} \tag{5.5}$$

It is easy to see that $\tau^* \in \widetilde{S}(\widetilde{H})$ implies $\pi_T(\tau^*) \in \widetilde{S}_T(\widetilde{H}_T)$ by the definition of $\widetilde{H}_T$. Hence, by (5.4), there exists a sequence $\{\tau_i'\}$ such that $\tau_i' \in S_T(\widetilde{H})$ and

$$\tau_i' \to \pi_T(\tau^*) \in \widetilde{S}_T(\widetilde{H}_T) \quad (i \to \infty). \tag{5.6}$$

Then, it follows from the claim 3. of Lemma 5.2 that there exists a sequence $\{\tau_i\}$ such that $\tau_i \in S(\widetilde{H})$ and $\pi_T(\tau_i) = \tau_i'$. Note that we cannot expect $\pi_{\{1,2,\cdots,m\}\setminus T}(\tau_i) \to 0$ $(i \to \infty)$. Hence we need to modify $\{\tau_i\}$ so that it converges to 0 with respect to the variables τ_k $(k \notin T)$.

We set, for a sequence $\{\lambda_i\}$ with $0 < \lambda_i < 1$,

$$\hat{\tau}_i = \mu_{j_1}(\cdots(\mu_{j_d}(\tau_i; \lambda_i); \cdots); \lambda_i).$$

Since $\mu_k(\tau; \lambda) \in S(\widetilde{H})$ holds if $0 < \lambda < 1$ and $\tau \in S(\widetilde{H})$, we have $\hat{\tau}_i \in S(\widetilde{H})$. Finally, by taking $\{\lambda_i\}$ to be rapidly decreasing to 0, we get the sequence $\{\hat{\tau}_i\}$ which satisfies

$$\hat{\tau}_i \in S(\widetilde{H}) \quad \text{and} \quad \hat{\tau}_i \to \tau^* \quad (i \to \infty)$$

thanks to (5.5) and (5.6). This completes the proof. □

Corollary 5.3. *Let H be a finite subset of $\mathcal{H}_\tau \times \mathbb{R}_{\geq 0}$ such that $[H]$ and $\mathcal{G}$ are equivalent and $Q(H) = H$ holds. For any $V \in V(\zeta)$ and for any $\epsilon_0 > 0$ and $\{\epsilon_{(f,v)}\}_{(f,v)\in H}$ with $\epsilon_{(f,v)} > 0$, the closure of the multi-cone*

$$S_H(V; \{\epsilon\}) = \left\{ z \in X; \begin{array}{ll} z \in V, \ |z^{(0)}| < \epsilon_0, & \\ f\left(|z^{(*)}|\right) < v + \epsilon_{(f,v)} & ((f,v) \in H) \end{array} \right\}$$

is given by

$$\left\{ z \in X; \begin{array}{l} z \in \overline{V},\ |z^{(0)}| \le \epsilon_0, \\ f_N\left(|z^{(*)}|\right) \le v f_D\left(|z^{(*)}|\right) \qquad ((f = f_N/f_D, v) \in \varpi(H_\epsilon)) \end{array} \right\}.$$

Remark 5.4. *In general, the closure of $S(V,\epsilon)$ is strictly smaller than the set defined by equalities where $<$ is replaced with $\le$. One can observe this fact for a clean intersection case: $Z_1 = \{z_1 = z_3 = 0\}$, $Z_2 = \{z_2 = z_3 = 0\}$, $Z_3 = \{z_1 = z_2 = z_3 = 0\}$. Then $S(V,\epsilon)$ is defined by*

$$\begin{cases} |z_3|/|z_1| & < \epsilon, \\ |z_3|/|z_2| & < \epsilon, \\ |z_1 z_2|/|z_3| & < \epsilon \end{cases}$$

and $z_k \in G_k$, where G_k is an open cone. Note that this set is relatively compact. Then,

$$\begin{cases} |z_3| & \le \epsilon |z_1|, \\ |z_3| & \le \epsilon |z_2|, \\ |z_1 z_2| & \le \epsilon |z_3| \end{cases}$$

with $z_k \in \overline{G_k}$ is not compact. In fact, the unbounded set $z_1 = z_3 = 0$ and $z_2 \in G_2$ is contained in this set. Therefore the set is strictly bigger than the closure of $S(V,\epsilon)$. To avoid this, we need to add some extra inequalities.

Theorem 5.5. *We have*

$$H^k(\rho^{-1}\nu_\chi^{sa}\mathcal{O}_X^{\mathrm{w}}) = 0 \qquad (k \neq 0). \tag{5.7}$$

Proof. In the proof, we follow the notations used in Subsection 4.1, and we may also assume from the beginning that all the entries of A_χ are non-negative integers and $\sigma_A = 1$. Let $p = (0;\zeta) = (0;\zeta^{(1)},\ldots,\zeta^{(m)}) \in S_\chi$. Define $\widetilde{\zeta} := (\widetilde{\zeta}^{(1)},\ldots,\widetilde{\zeta}^{(m)})$ by $\widetilde{\zeta}^{(k)} = \zeta^{(k)}/|\zeta^{(k)}|$ $(k \notin J_Z(\zeta))$ and $\widetilde{\zeta}^{(k)} = 0$ otherwise. By the observations in this section, we have

$$(\rho^{-1}H^k\nu_\chi^{sa}\mathcal{O}_X^{\mathrm{w}})_p \simeq \varinjlim_{V,\epsilon} H^k(X; \mathbb{C}_{\overline{S(V,\epsilon)}} \overset{\mathrm{w}}{\otimes} \mathcal{O}_X),$$

where V runs through the family $V(\zeta)$ and $\epsilon > 0$. As the closed subset $\overline{S(V,\epsilon)}$ does not satisfy the "property (λ)" of [6] in general, we need to replace the family $\{\overline{S(V,\epsilon)}\}_{V,\epsilon}$ with an equivalent family of closed analytic polyhedra which, in particular, satisfy the "property (λ)". Then the result immediately follows from the main theorem in [6].

Let us construct an equivalent family. We may assume, in what follows, the cone V is always taken to be an analytic polyhedron. Set

$$\check{N} := \{\alpha = (\alpha_1, \dots, \alpha_m) \in \mathbb{Z}^m_{\geq 0};\ \alpha_k = 0\ (k \in J_Z(\zeta))\}$$

and, for $\alpha = (\alpha_1, \dots, \alpha_m) \in \mathbb{Z}^m_{\geq 0}$,

$$h_\alpha(z) := \prod_{1 \leq k \leq m} |z^{(k)}|^{\alpha_k}.$$

We also denote by Θ the set of strictly increasing continuous functions $\theta(t)$ on $(0, +\infty)$ satisfying $\lim_{t \to 0+0} \theta(t) = 0$. By Corollary 5.3 on the closure of $S(V, \epsilon)$, we may assume that each closed subset $\overline{S(V, \epsilon)}$ in the family $\{\overline{S(V, \epsilon)}\}_{V,\epsilon}$ has the following form:

$$\left(\bigcap_{(\alpha,\beta,\theta) \in P} \{z \in \overline{V};\ h_\alpha(z) \leq \theta(\epsilon) h_\beta(z)\} \right) \bigcap$$
$$\left(\bigcap_{(\gamma,\delta,v,\theta) \in Q} \{z \in \overline{V};\ h_\gamma(z) \leq (v + \theta(\epsilon)) h_\delta(z)\} \right) \bigcap \left\{ |z^{(0)}| \leq \epsilon \right\},$$

where P is a finite subset of $(\mathbb{Z}^m_{\geq 0}) \times \check{N} \times \Theta$ and Q is also a finite subset of $\check{N} \times \check{N} \times \mathbb{R}_{>0} \times \Theta$. Note that α belongs to $\mathbb{Z}^m_{\geq 0}$, however, the other β, δ and γ belong to $\check{N}$, which is crucial in the subsequent arguments.

Define the family $\{H(V, \epsilon)\}_{V,\epsilon>0}$ by

$$\{z \in \overline{V};\ h_\alpha(z) \leq \theta(\epsilon) h_\beta(z)\}$$

with a fixed $(\alpha, \beta, \theta) \in (\mathbb{Z}^m_{\geq 0}) \times \check{N} \times \Theta$, and define also the family $\{G(V, \epsilon)\}_{V,\epsilon>0}$ by

$$\{z \in \overline{V};\ h_\gamma(z) \leq (v + \theta(\epsilon)) h_\delta(z)\}$$

with a fixed $(\gamma, \delta, v, \theta) \in \check{N} \times \check{N} \times \mathbb{R}_{>0} \times \Theta$.

First we will make an equivalent family $\{\widetilde{H}(\Xi, V, \epsilon)\}_{\Xi,V,\epsilon}$ to $\{H(V, \epsilon)\}_{V,\epsilon>0}$ which consists of closed analytic polyhedra as follows: For a vector

$$\xi = (\xi^{(1)}, \dots, \xi^{(m)}) \in \mathbb{C}^{n_1} \times \cdots \times \mathbb{C}^{n_m}$$

with $|\xi^{(k)}| \neq 0$ $(1 \leq k \leq m)$, and for $\alpha \in \mathbb{Z}^m_{\geq 0}$, set

$$\widehat{h}_{\xi,\alpha}(z) := \operatorname{Re} \prod_{1 \leq k \leq m} \left(\frac{|\xi^{(k)}| \left\langle z^{(k)}, \overline{\xi^{(k)}} \right\rangle}{\left\langle \xi^{(k)}, \overline{\xi^{(k)}} \right\rangle} \right)^{\alpha_k},$$

where $\langle a, b \rangle := a_1 b_1 + \cdots + a_r b_r$ for $a = (a_1, \dots, a_r)$ and $b = (b_1, \dots, b_r)$ and $\overline{a}$ denotes the complex conjugate of a complex vector a. Then we can easily confirm the following facts: For any $z \in X$, we have

$$\widehat{h}_{\xi,\alpha}(z) \leq (\sqrt{n})^{|\alpha|} h_\alpha(z). \tag{5.8}$$

Furthermore, for $\beta \in \check{N}$ and $V \in V(\zeta) = V(\widetilde{\zeta})$, there exists constant $\kappa > 1$ satisfying

$$\kappa^{-1} h_\beta(z) \le \widehat{h}_{\widetilde{\zeta},\beta}(z) \le \kappa h_\beta(z) \qquad (z \in \overline{V}). \tag{5.9}$$

Note that, here, if V becomes smaller as a cone, then we can take the above κ to be closer to 1.

Let $\Xi = \{\Xi_k\}_{k \in J_Z(\zeta)}$ be a family of the sets of vectors where each set Ξ_k ($k \in J_Z(\zeta)$) consists of finitely many vectors $\xi^{(k)}$ in $\mathbb{C}^{n_k}$ with $|\xi^{(k)}| = 1$. Then we define the closed analytic polyhedra $\widetilde{H}(\Xi, V, \epsilon)$ by

$$\bigcap_{\xi} \left\{ z \in \overline{V};\ \widehat{h}_{\xi,\alpha}(z) \le \epsilon \widehat{h}_{\widetilde{\zeta},\beta}(z) \right\},$$

where the vector $\xi = (\xi^{(1)}, \ldots, \xi^{(m)})$ runs through the ones satisfying $\xi^{(k)} = \widetilde{\zeta}^{(k)}$ for $k \notin J_Z(\zeta)$ and $\xi^{(k)} \in \Xi_k$ for $k \in J_Z(\zeta)$. We can easily confirm that, by (5.8) and (5.9), the family $\{\widetilde{H}(\Xi, V, \epsilon)\}_{\Xi, V, \epsilon}$ and the one $\{H(V, \epsilon)\}_{V,\epsilon}$ are equivalent.

For an equivalent family to the family $\{G(V, \epsilon)\}_{V,\epsilon>0}$, by noticing the remark after (5.9), we can construct an equivalent family $\{\widetilde{G}(V, \epsilon)\}_{V,\epsilon}$ of $\{G(V, \epsilon)\}_{V,\epsilon>0}$ consisting of closed analytic polyhedra by

$$\widetilde{G}(V, \epsilon) := \left\{ z \in \overline{V};\ \widehat{h}_{\widetilde{\zeta},\gamma}(z) \le (v + \epsilon) \widehat{h}_{\widetilde{\zeta},\delta}(z) \right\}.$$

Hence we have obtained a required equivalent family of $\{\overline{S(V, \epsilon)}\}_{V,\epsilon}$ and the theorem follows. □

Theorem 5.6. *Let $p \in S_\chi$ be a point outside fixed points. Assume that the associated action μ is non-degenerate. Then we have*

$$H^k(\rho^{-1} \nu_\chi^{sa} \mathcal{O}^{\mathrm{w}}_{X|X \setminus \bigcup_k Z})_p = 0 \qquad (k \ne 0).$$

Let $p = (0; \zeta) = (0; \zeta^{(1)}, \ldots, \zeta^{(m)})$. Let φ_k ($1 \le k \le m$) be the monomials associated with the action μ. We may assume from the beginning that all the entries of A_χ are non-negative integers and $\sigma_A = 1$. Since μ is non-degenerate and p is outside fixed points, we have Rank $A_\chi = \ell \le m$ and we may assume that the first $\ell \times \ell$ sub-matrix in A_χ is invertible and $\zeta^{(k)} \ne 0$ holds for $1 \le k \le \ell$.

Now we prove this theorem after some preparations. We regard

$$w = (w_1, \ldots, w_\ell) = \varphi(t) = (\varphi_1(t), \ldots, \varphi_\ell(t))$$

as a holomorphic map from $\mathbb{C}^\ell_t$ to $\mathbb{C}^\ell_w$. Set

$$D := \{w_1 w_2 \cdots w_\ell = 0\} \subset \mathbb{C}^n_w \text{ and } L := \{t_1 \cdots t_\ell = 0\} \subset \mathbb{C}^n_t.$$

Then it is easy to see

$$\varphi^{-1}(D) = L.$$

Furthermore, as $w = \varphi(t)$ has the inverse $t = \varphi^{-1}(w)$ consisting of monomials of rational powers, the map

$$\varphi|_{\mathbb{C}^n_t \setminus L} : (\mathbb{C}^n_t \setminus L) \longrightarrow (\mathbb{C}^n_w \setminus D)$$

is a finite covering.

Let Ξ be the set of open poly-sectors $V = V_1 \times V_2 \times \cdots \times V_\ell$ in $\mathbb{C}^\ell$ where each V_k is an open proper sector in $\mathbb{C}$ and $1 \in V_k$. We denote by $\hat{A}_\chi$ the sub-matrix of A_χ consisting of the leftmost ℓ columns which is, by the assumption, invertible. Then, as

$$\arg(w) = (\arg(w_1), \ldots, \arg(w_\ell)) = \arg(t)\hat{A}_\chi \qquad (t \in \mathbb{C}^\ell \setminus L), \tag{5.10}$$

holds, we have the following:

For any $V \in \Xi$, there exists $V', W \in \Xi$ such that $V' \subset \varphi(W) \subset V$.

Remember that n_k is the dimension of the k-th coordinates block. Now, by a linear coordinates transformation on each coordinate block, we may assume that each direction $\zeta^{(k)}$ has a form

$$(|\zeta^{(k)}|, 0, \ldots, 0) \in \mathbb{C}^{n_k} \qquad (1 \le k \le m).$$

Furthermore we may assume $|\zeta^{(k)}| = 1$ for $1 \le k \le \ell$ by considering normalization of ζ. Set

$$\begin{array}{ccccccccc} Y = & \mathbb{C}^{n_0} & \times & \left(\underset{1\le k\le \ell}{\times} \mathbb{C}^{n_k-1}\right) & \times & \left(\underset{\ell<k\le m}{\times} \mathbb{C}^{n_k}\right) & \times & \mathbb{C}^\ell. \\ & \big(z^{(0)} & , & z_*^{(1)}, \ldots, z_*^{(\ell)} & , & z^{(\ell+1)}, \cdots, z^{(m)} & , & t\big). \end{array}$$

Then define the map $f : Y \to X$ by

$$\begin{aligned} &\big(z^{(0)},\, z_*^{(1)}, \ldots, z_*^{(\ell)},\, z^{(\ell+1)}, \cdots, z^{(m)},\, t\big) \\ &\quad \longrightarrow \big(z^{(0)},\, \varphi_1(t)(1, z_*^{(1)}), \ldots, \varphi_\ell(t)(1, z_*^{(\ell)}),\, \varphi_{\ell+1}(t) z^{(\ell+1)}, \cdots, \varphi_m(t) z^{(m)}\big). \end{aligned}$$

We denote by $z_1^{(k)}$ the first coordinate of the coordinates block $z^{(k)}$ and by $z_*^{(k)}$ the rest, i.e., $z^{(k)} = (z_1^{(k)}, z_*^{(k)})$. Define the subset in X by

$$T_X := \{z_1^{(1)} z_1^{(2)} \cdots z_1^{(\ell)} = 0\} \subset X$$

and the one in Y by

$$T_Y := \{t_1 t_2 \cdots t_\ell = 0\} \subset Y.$$

By the previous observations and by noticing the fact $\varphi_k(t) \neq 0$ if $t_1 \ldots t_\ell \neq 0$, we have

$$f^{-1}(T_X) = T_Y$$

and

$$\text{the map } f|_{Y \setminus T_Y} \text{ is a finite covering over } X \setminus T_X.$$

Furthermore, we have

$$Z_1 \cup Z_2 \cup \cdots \cup Z_\ell = f(T_Y) \subset T_X.$$

Let $\epsilon_Y > 0$ and $V_Y \in \Xi$. Then define the open subset $S_Y(V_Y, \epsilon_Y)$ in Y by

$$\left\{ \begin{array}{l} |z^{(0)}| < \epsilon_Y, \ |z_*^{(k)}| < \epsilon_Y \ (1 \le k \le \ell), \\ |z^{(k)} - \zeta^{(k)}| < \epsilon_Y \ (\ell < k \le m), \\ t \in V_Y, \ |t_k| < \epsilon_Y \ (1 \le k \le \ell) \end{array} \right\}.$$

We first show the following lemma:

Lemma 5.7. *For a multi-cone $S_X(V_X, \epsilon_X) \subset X$ with $\epsilon_X > 0$ and $V_X \in V(\zeta)$, we can find some $S_Y(V_Y, \epsilon_Y)$ with $\epsilon_Y > 0$ and $V_Y \in \Xi$ such that*

$$f(S_Y(V_Y, \epsilon_Y)) \subset S_X(V_X, \epsilon_X).$$

Proof. In fact, if ϵ_Y is sufficiently small and V_Y is sufficiently thin, then clearly $f(S_Y(V_Y, \epsilon_Y)) \subset V_X$ holds. As we have

$$\varphi_k^{-1}(\varphi_1(t)(1, z_*^{(1)}), \ldots, \varphi_\ell(t)(1, z_*^{(\ell)})) = t_k \varphi_k^{-1}((1, z_*^{(1)}), \ldots, (1, z_*^{(\ell)})),$$

and as $|\varphi_k^{-1}((1, z_*^{(1)}), \ldots, (1, z_*^{(\ell)}))|$ is closed to 1, we see that $|\varphi_k^{-1}| < \epsilon_X$ holds at a point in $f(S_Y(V_Y, \epsilon_Y))$. In the same way, we have

$$\begin{aligned} &\psi_k(\varphi_1(t)(1, z_*^{(1)}), \ldots, \varphi_\ell(t)(1, z_*^{(\ell)}), \varphi_{\ell+1}(t) z^{(\ell+1)}, \cdots, \varphi_m(t) z^{(m)}) \\ &\quad = z^{(k)} / \varphi_k(\varphi^{-1}((1, z_*^{(1)}), \ldots, (1, z_*^{(\ell)})). \end{aligned}$$

Since $|\varphi_k(\varphi^{-1}((1, z_*^{(1)}), \ldots, (1, z_*^{(\ell)}))|$ is very close to 1 and $|z^{(k)}|$ is nearly $|\zeta^{(k)}|$, we know that

$$|\zeta^{(k)}| - \epsilon_X < |\psi_k| < |\zeta^{(k)}| + \epsilon_X$$

holds at a point in $f(S_Y(V_Y, \epsilon_Y))$. Hence we have obtained the desired inclusion. □

On the other hand, we have the inverse of f defined on a fixed $S_X(V_X, \epsilon_X)$ ($\epsilon_X > 0$ and $V_X \in V(\zeta)$), which is explicitly given by

$$\begin{aligned} &\left(z^{(0)}, \ z^{(1)}, \ldots, z^{(m)}\right) \in X \\ &\quad \longrightarrow \left(z^{(0)}, \ z_*^{(1)}/z_1^{(1)}, \ldots, z_*^{(\ell)}/z_1^{(\ell)}, \ z^{(\ell+1)}/\varphi_{\ell+1}(t), \cdots, z^{(m)}/\varphi_m(t), \ t\right) \in Y, \end{aligned}$$

where t is determined by

$$t_k = \varphi_k^{-1}(z_1^{(1)}, z_1^{(2)}, \ldots, z_1^{(\ell)}) \quad (1 \le k \le \ell). \tag{5.11}$$

Lemma 5.8. *For a given* $S_Y(V_Y, \epsilon_Y)$ *with* $\epsilon_Y > 0$ *and* $V_Y \in \Xi$, *we can find* $S_X(V'_X, \epsilon'_X)$ *with* $0 < \epsilon'_X < \epsilon_X$ *and* $V'_X \subset V_X$ *in* $V(\zeta)$ *such that*

$$f^{-1}(S_X(V'_X, \epsilon'_X)) \subset S_Y(V_Y, \epsilon_Y).$$

Proof. This is shown by the same argument as the one in the previous lemma. First we note that, for any $\delta > 0$, if we take $V'_X \in V(\zeta)$ sufficiently thin, then we have

$$(1-\delta)|z_1^{(k)}| \le |z^{(k)}| \le (1+\delta)|z_1^{(k)}| \qquad (1 \le k \le m) \tag{5.12}$$

hold on this V'_X.

It follows from (5.10) and (5.11) that we have $t \in V_Y$ if ϵ'_X is sufficiently small and V'_X is thin. Furthermore, $z \in S_X(V'_X, \epsilon'_X)$ implies $\varphi_k^{-1}(|z^{(*)}|) < \epsilon'_X$ for $1 \le k \le \ell$. Hence, by noticing (5.12), we get $|t_k| < \epsilon_Y$.

For $1 \le k \le \ell$, as $z^{(k)}$ is a point in a cone in $\mathbb{C}^{n_k}$ with direction $(1, 0, \dots, 0)$, by (5.12), we have

$$|z_*^{(k)}/z_1^{(k)}| < \epsilon_Y \qquad (1 \le k \le \ell).$$

Since $|\zeta^{(k)}| - \epsilon'_X < \psi_k(|z^{(*)}|) < |\zeta^{(k)}| + \epsilon'_X$ $(\ell < k \le m)$ follows from $z \in S_X(V'_X, \epsilon'_X)$, by taking (5.12) into account, for any $\delta' > 0$, we get

$$(1-\delta')|\zeta^{(k)}| < |z^{(k)}/t_k| < (1+\delta')|\zeta^{(k)}| \qquad (\ell < k \le m)$$

if $\epsilon'_X > 0$ is taken to be so small and V'_X is so thin. Hence, since $z \in V'_X$, in particular, $z^{(k)}$ belongs to a thin cone in $\mathbb{C}^{n_k}$ with direction $\zeta^{(k)}$ and since $\arg(t_k)$ is nearly 0, we conclude

$$|z^{(k)}/t_k - \zeta^{(k)}| < \epsilon_Y.$$

This completes the proof. □

For any $V_Y \in \Xi$, since $Z_1 \cup \cdots \cup Z_\ell = f(T_Y) \subset T_X$ and f is locally isomorphic outside T_Y, we have

$$\overline{f\left((\mathbb{C}^{n-\ell} \times V_Y) \cap U\right)} \cap T_X = \overline{f\left((\mathbb{C}^{n-\ell} \times V_Y) \cap U\right)} \cap (Z_1 \cup Z_2 \cup \dots Z_\ell)$$

for a bounded open subset U in Y. Therefore we have

$$\begin{aligned} f\left(\overline{S_Y(V_Y, \epsilon_Y)} \setminus T_Y\right) &= f\left(\overline{S_Y(V_Y, \epsilon_Y)}\right) \setminus T_X \\ &= \overline{f\left(S_Y(V_Y, \epsilon_Y)\right)} \setminus T_X \\ &= \overline{f(S_Y(V_Y, \epsilon_Y))} \setminus (Z_1 \cup \cdots \cup Z_\ell). \end{aligned}$$

Set $S_Y := S_Y(V_Y, \epsilon_Y)$ for short. Then, as $f|_{\overline{S_Y} \setminus T_Y}$ is an isomorphism onto its image, we have

$$Rf_! \mathbb{C}_{\overline{S_Y} \setminus T_Y} = \mathbb{C}_{f(\overline{S_Y} \setminus T_Y)} = \mathbb{C}_{\overline{f(S_Y)} \setminus (Z_1 \cup \cdots \cup Z_\ell)}.$$

Hence it follows from Theorem 5.7 of [15] that we have

$$R\Gamma(Y; R\mathcal{H}om_{\mathcal{D}_Y}(\mathcal{D}_{Y \xrightarrow{f} X}, \mathbb{C}_{\overline{S_Y} \setminus T_Y} \overset{\mathrm{w}}{\otimes} \mathcal{O}_Y) \simeq R\Gamma(X; \mathbb{C}_{\overline{f(S_Y)} \setminus (Z_1 \cup \cdots \cup Z_\ell)} \overset{\mathrm{w}}{\otimes} \mathcal{O}_X). \quad (5.13)$$

Since f is locally isomorphic outside T_Y, its Jacobian J_f does not vanish outside T_Y. Therefore, for any point $p \in Y$, J_f has the form $g\, t_1^{\alpha_1} t_2^{\alpha_2} \cdots t_\ell^{\alpha_\ell}$ for some multi-index $\alpha = (\alpha_1, \dots, \alpha_\ell) \in \mathbb{Z}_{\geq 0}^\ell$ and some holomorphic function g with $g(p) \neq 0$.

Let $h := t_1 t_2 \cdots t_\ell$, and let $\mathcal{O}_{Y,h}$ denote the sheaf of meromorphic functions whose poles are contained in T_Y. (i.e., $\mathcal{O}_{Y,h} = \mathcal{O}_Y[t_1^{-1}, \dots, t_\ell^{-1}]$). We set $\mathcal{D}_{Y,h} := \mathcal{O}_{Y,h} \underset{\mathcal{O}_Y}{\otimes} \mathcal{D}_Y$. Since J_f is invertible in $\mathcal{O}_{Y,h}$, we have

$$\mathcal{D}_{Y,h} \underset{\mathcal{D}_Y}{\otimes} \mathcal{D}_{Y \xrightarrow{f} X} \simeq \mathcal{D}_{Y,h}. \quad (5.14)$$

(See p. 479 in [9] also). As $\mathbb{C}_{\overline{S_Y} \setminus T_Y} \overset{\mathrm{w}}{\otimes} \mathcal{O}_Y$ is a $\mathcal{D}_{Y,h}$-module and $\mathcal{D}_{Y,h}$ is flat over $\mathcal{D}_Y$, we have

$$R\mathcal{H}om_{\mathcal{D}_Y}(\mathcal{D}_{Y \xrightarrow{f} X}, \mathbb{C}_{\overline{S_Y} \setminus T_Y} \overset{\mathrm{w}}{\otimes} \mathcal{O}_Y) \simeq R\mathcal{H}om_{\mathcal{D}_{Y,h}}(\mathcal{D}_{Y,h} \underset{\mathcal{D}_Y}{\otimes} \mathcal{D}_{Y \xrightarrow{f} X}, \mathbb{C}_{\overline{S_Y} \setminus T_Y} \overset{\mathrm{w}}{\otimes} \mathcal{O}_Y),$$

and hence, by (5.13) and (5.14), we finally obtain

$$R\Gamma(Y; \mathbb{C}_{\overline{S_Y} \setminus T_Y} \overset{\mathrm{w}}{\otimes} \mathcal{O}_Y) \simeq R\Gamma(X; \mathbb{C}_{\overline{f(S_Y)} \setminus (Z_1 \cup \cdots \cup Z_\ell)} \overset{\mathrm{w}}{\otimes} \mathcal{O}_X). \quad (5.15)$$

Now we are ready to give the proof of the theorem.

Proof. By the stalk formula, it suffices to show

$$\begin{aligned} &H^k(\rho^{-1}\nu_X^{sa}(\mathcal{O}^{\mathrm{w}}_{X|X\setminus \bigcup_k Z}))_p \\ &= \varinjlim_{\epsilon_X > 0,\, V_X \in V(\zeta)} H^k(S_X(V_X, \epsilon_X); \mathcal{O}^{\mathrm{w}}_{X|X\setminus \bigcup_k Z}) \\ &= \varinjlim_{\epsilon_X > 0,\, V_X \in V(\zeta)} H^k(X; \mathbb{C}_{\overline{S_X(V_X, \epsilon_X)} \setminus (Z_1 \cup \cdots \cup Z_\ell)} \overset{\mathrm{w}}{\otimes} \mathcal{O}_X) = 0 \quad (k \neq 0). \end{aligned}$$

By Lemmas 5.7 and 5.8 and (5.15), the claim is equivalent to

$$\varinjlim_{\epsilon_Y > 0,\, V_Y \in \Xi} H^k(Y; \mathbb{C}_{\overline{S_Y(V_Y, \epsilon_Y)} \setminus T_Y} \overset{\mathrm{w}}{\otimes} \mathcal{O}_Y) = 0 \quad (k \neq 0).$$

Recall that V_Y has a form $V_1 \times \cdots \times V_\ell$ where each V_k is a proper open sector in $\mathbb{C}$ containing 1. Since we have

$$\overline{S_Y(V_Y, \epsilon_Y)} \setminus T_Y = B \times (\overline{V_1} \setminus \{0\}) \times \cdots \times (\overline{V_\ell} \setminus \{0\}),$$

where B is a closed convex set in $\mathbb{C}^{n-\ell}$, it follows from the topological tensor product formula in [15] that we get

$$\begin{aligned} &R\Gamma(Y; \mathbb{C}_{\overline{S_Y(V_Y, \epsilon_Y)} \setminus T_Y} \overset{\mathrm{w}}{\otimes} \mathcal{O}_Y) \\ &= R\Gamma(\mathbb{C}^{n-\ell}; \mathbb{C}_B \overset{\mathrm{w}}{\otimes} \mathcal{O}) \widehat{\boxtimes} R\Gamma(\mathbb{C}; \mathbb{C}_{\overline{V_1} \setminus \{0\}} \overset{\mathrm{w}}{\otimes} \mathcal{O}) \widehat{\boxtimes} \cdots \widehat{\boxtimes} R\Gamma(\mathbb{C}; \mathbb{C}_{\overline{V_\ell} \setminus \{0\}} \overset{\mathrm{w}}{\otimes} \mathcal{O}). \end{aligned}$$

As B is a convex closed set, $R\Gamma(\mathbb{C}^{n-\ell}; \mathbb{C}_B \overset{\mathrm{w}}{\otimes} \mathcal{O})$ is concentrated in degree 0 and the 0-th cohomology group is naturally equipped with **FN** topology. For $1 \leq k \leq \ell$, $R\Gamma(\mathbb{C}; \mathbb{C}_{\overline{V_k} \setminus \{0\}} \overset{\mathrm{w}}{\otimes} \mathcal{O})$ is also known to be concentrated in degree 0 (see, for example, Lemma 8.5 [9]) and the 0-th cohomology group has **FN** topology. Hence we can conclude that $R\Gamma(Y; \mathbb{C}_{\overline{S_Y(V_Y, \epsilon_Y)} \setminus T_Y} \overset{\mathrm{w}}{\otimes} \mathcal{O}_Y)$ is concentrated in degree 0. This completes the proof. □

Thanks to Theorems 5.5 and 5.6 we obtain the following result

Theorem 5.9. *Assume the associated action μ is non-degenerate. Then the distinguished triangle* (4.24) *induces an exact sequence (outside the fixed points)*

$$0 \to \rho^{-1} H^0 \nu_\chi^{sa} \mathcal{O}^{\mathrm{w}}_{X|X \setminus \bigcup_k Z} \to \rho^{-1} H^0 \nu_\chi^{sa} \mathcal{O}^{\mathrm{w}}_X \to \rho^{-1} H^0 \nu_\chi^{sa} \mathcal{O}^{\mathrm{w}}_{X|\bigcup_k Z} \to 0. \tag{5.16}$$

All the complexes $\rho^{-1} \nu_\chi^{sa} \mathcal{O}^{\mathrm{w}}_{X|X \setminus \bigcup_k Z}$, $\rho^{-1} \nu_\chi^{sa} \mathcal{O}^{\mathrm{w}}_X$ and $\rho^{-1} \nu_\chi^{sa} \mathcal{O}^{\mathrm{w}}_{X|\bigcup_k Z}$ are concentrated in degree zero.

Note that, by Theorem 5.5, the complex $\rho^{-1} \nu_\chi^{sa} \mathcal{O}^{\mathrm{w}}_X$ is always concentrated in degree zero even if the action μ is degenerate and the point p is in the fixed points.

Let us summarize our previous results. We assume $X = \mathbb{C}^n$. Recall that the set of points in S_χ outside the fixed points is denoted by S°_χ. Let W be an $(\mathbb{R}^+)^\ell$-conic subanalytic open subset in S°_χ. Let us consider the multi-specialization of Whitney holomorphic functions. We have

$$\Gamma(W; \rho^{-1} H^0 \nu_\chi^{sa} \mathcal{O}^{\mathrm{w}}_{X|X \setminus \bigcup_k Z}) \simeq \varprojlim_{V'} \varinjlim_{U'} \Gamma(U'; \mathcal{O}^{\mathrm{w}}_{X|X \setminus \bigcup_k Z}),$$

$$\Gamma(W; \rho^{-1} H^0 \nu_\chi^{sa} \mathcal{O}^{\mathrm{w}}_X) \simeq \varprojlim_{V'} \varinjlim_{U'} \Gamma(U'; \mathcal{O}^{\mathrm{w}}_X),$$

$$\Gamma(W; \rho^{-1} H^0 \nu_\chi^{sa} \mathcal{O}^{\mathrm{w}}_{X|\bigcup_k Z}) \simeq \varprojlim_{V'} \varinjlim_{U'} \Gamma(U'; \mathcal{O}^{\mathrm{w}}_{X|\bigcup_k Z}),$$

where V' ranges through a family of $(\mathbb{R}^+)^\ell$-conic subanalytic open subsets which are compactly generated in W, U' ranges through the family of $\mathrm{Op}(X_{sa})$ such that $C_\chi(X \setminus U') \cap V' = \emptyset$. Here we say that V' is compactly generated in W if there exists a compact subset $K \subset W$ with $V' \subset (\mathbb{R}^+)^\ell K$.

Recall that the classical sheaf $\mathcal{A}^{<0}_\chi$ (resp. $\mathcal{A}_\chi$, resp. $\mathcal{A}^{CF}_\chi$) on S°_χ was defined in Subsection 3.7 in [10]. For $\{S_p\}_{p \in W} \in \mathcal{S}(W)$ (for these definitions, see Subsection 3.7 [10]) and a compactly generated $(\mathbb{R}^+)^\ell$-conic subset V' in W, we can find finitely many points $P \subset W$ such that

$$C_\chi\left(X \setminus (\bigcup_{p \in P} S_p)\right) \cap V' = \emptyset.$$

Hence, by Theorems 3.37, 3.45 and 3.49 in [10] the identity morphism induces morphisms of presheaves on $S_{\mathcal{X}}^{\circ}$

$$\begin{aligned}
\widetilde{\mathcal{A}}_{\mathcal{X}}^{<0} &\to \rho^{-1}H^0\nu_{\mathcal{X}}^{sa}\mathcal{O}_{X|X\setminus\bigcup_k Z}^{\mathrm{w}},\\
\widetilde{\mathcal{A}}_{\mathcal{X}} &\to \rho^{-1}H^0\nu_{\mathcal{X}}^{sa}\mathcal{O}_{X}^{\mathrm{w}},\\
\widetilde{\mathcal{A}}_{\mathcal{X}}^{CF} &\to \rho^{-1}H^0\nu_{\mathcal{X}}^{sa}\mathcal{O}_{X|\bigcup_k Z}^{\mathrm{w}}.
\end{aligned}$$

Since the above morphisms are isomorphisms in the stalks, we have obtained the isomorphisms of the sheaves on $S_{\mathcal{X}}^{\circ}$

$$\begin{aligned}
\mathcal{A}_{\mathcal{X}}^{<0} &\simeq \rho^{-1}H^0\nu_{\mathcal{X}}^{sa}\mathcal{O}_{X|X\setminus\bigcup_k Z}^{\mathrm{w}},\\
\mathcal{A}_{\mathcal{X}} &\simeq \rho^{-1}H^0\nu_{\mathcal{X}}^{sa}\mathcal{O}_{X}^{\mathrm{w}},\\
\mathcal{A}_{\mathcal{X}}^{CF} &\simeq \rho^{-1}H^0\nu_{\mathcal{X}}^{sa}\mathcal{O}_{X|\bigcup_k Z}^{\mathrm{w}}.
\end{aligned}$$

By Proposition 5.9, on $S_{\mathcal{X}}^{\circ}$, we have the exact sequence of sheaves

$$0 \to \mathcal{A}_{\mathcal{X}}^{<0} \to \mathcal{A}_{\mathcal{X}} \to \mathcal{A}_{\mathcal{X}}^{CF} \to 0, \tag{5.17}$$

when the associated action μ is non-degenerate. The above exact sequence (5.17) is a generalization of the Borel-Ritt exact sequence for multi-asymptotically developable functions.

5.2 Vanishing theorems for tempered holomorphic functions

We are now going to consider the vanishing of the cohomology for the multi-specialization of the sheaf of tempered holomorphic functions of [16]. Let X be a real analytic manifold ($X = \mathbb{R}^n$ in the local model). Let $\mathcal{D}b_X$ denote the sheaf of distributions on X.

Definition 5.10. *One denotes by $\mathcal{D}b_X^t$ the presheaf of tempered distributions on X_{sa} defined as follows:*

$$U \mapsto \Gamma(X;\mathcal{D}b_X)/\Gamma_{M\setminus U}(X;\mathcal{D}b_X).$$

As a consequence of the Łojasievicz's inequalities [17], for $U, V \in \mathrm{Op}(X_{sa})$ the sequence

$$0 \to \mathcal{D}b_X^t(U\cup V) \to \mathcal{D}b_X^t(U)\oplus\mathcal{D}b_X^t(V) \to \mathcal{D}b_X^t(U\cap V) \to 0$$

is exact. This implies that $\mathcal{D}b_X^t$ is a sheaf on X_{sa}. Moreover it follows by definition that $\mathcal{D}b_X^t$ is quasi-injective.

Let X be a complex manifold ($X = \mathbb{C}^n$ in the local model) and let $X_{\mathbb{R}}$ denote the underlying real analytic manifold of X. One denotes by $\mathcal{O}_X^t$ the sheaf defined as follows:

$$\mathcal{O}_X^t := R\mathcal{H}om_{\rho_!\mathcal{D}_{\overline{X}}}(\mathcal{O}_{\overline{X}}, \mathcal{D}b_{X_{\mathbb{R}}}^t).$$

Let U be an open subset in X. Then $H^0(U; \mathcal{O}_X^t)$ consists of tempered distributions on U that satisfy the Cauchy-Riemann system.

Let $\chi = \{Z_1, \ldots, Z_\ell\}$ be a family of complex submanifolds of X. Theorem 5.5 holds for $\mathcal{O}_X^t$ and $\mathcal{O}_X$ (the sheaf of holomorphic functions).

Theorem 5.11. *Under the same geometrical situation as in Theorem 5.5, we have*

$$H^k(\rho^{-1}\nu_\chi^{sa}\mathcal{O}_X^t) = 0 \qquad (k \neq 0), \tag{5.18}$$

and

$$H^k(\nu_\chi \mathcal{O}_X) = 0 \qquad (k \neq 0). \tag{5.19}$$

Proof. We may assume that all the entries of A_χ are non-negative integers and $\sigma_A = 1$. Let $p = (0; \zeta) = (0; \zeta^{(1)}, \cdots, \zeta^{(m)}) \in S_\chi$ and H a finite subset in $\mathcal{H}_\tau \times \mathbb{R}_{\geq 0}$ for which $[H]$ and $\mathcal{G}$ are equivalent. It follows from the stalk formula that

$$(\rho^{-1}H^k\nu_\chi^{sa}\mathcal{O}_X^t)_p \simeq \varinjlim_{V, \{\epsilon\}} H^k(S_H(V, \{\epsilon\}); \mathcal{O}_X^t),$$

where V runs through $V(\zeta)$ and $\{\epsilon\}$ ranges through families of positive real numbers. We may additionally assume that H satisfies $H = Q(H)$, and thus, a multi-cone $S_H(V, \{\epsilon\})$ has the form

$$\left\{ z \in X; \begin{array}{l} z \in V, \ |z^{(0)}| < \epsilon_0, \\ f\left(|z^{(*)}|\right) < v + \epsilon_{(f,v),+} \text{ for any } (f, v) \in H \end{array} \right\} \tag{5.20}$$

for a convex cone $V \in V(\zeta)$ and a family $\{\epsilon\}$ of positive real numbers.

Set

$$Y = \{\eta = (\eta_1, \ldots, \eta_m) \in \mathbb{C}^m; \ \eta_k \neq 0 \ (k \notin J_Z(\zeta))\}.$$

Then, for each $(f, v) \in H$, $f(\eta)$ is a holomorphic function on Y as $\nu_k(f) \geq 0$ holds for $k \in J_Z(\zeta)$. Let us consider the set

$$\begin{aligned} Y_H &:= \{\eta \in Y; \ |f(\eta)| < v + \epsilon_{(f,v),+} \text{ for any } (f, v) \in H\} \\ &= \bigcap_{(f,v) \in H} f^{-1}(D_{v+\epsilon_{(f,v),+}}). \end{aligned}$$

Here $D_\delta \subset \mathbb{C}$ denotes the open disk with radius $\delta > 0$ and center at the origin, and each f is regarded as a holomorphic map from Y to $\mathbb{C}$. Then Y_H becomes a Stein open subset because the inverse image of a Stein subset by a holomorphic map on a Stein subset is again Stein and Y itself is a Stein subset in $\mathbb{C}^m$. Let Θ be the set of morphisms

$$p(z) = (p_1(z), \ldots, p_m(z)) \in \mathbb{C}^m, \qquad z = (z_1, \ldots, z_n) \in \mathbb{C}^n$$

from $\mathbb{C}^n$ to $\mathbb{C}^m$ such that, if $k \in J_Z(\zeta)$, $p_k(z) = z_{i_k}$ with $z_{i_k} \in z^{(k)}$ (i.e., z_{i_k} is one of the coordinate variables in $z^{(k)}$) and, for $k \notin J_Z(\zeta)$,

$$p_k(z) = c_k \langle z, \overline{\zeta^{(k)}} \rangle$$

with the constant $c_k = |\zeta^{(k)}| / \langle \zeta^{(k)}, \overline{\zeta^{(k)}} \rangle$. Note that Θ is a finite set. We define

$$\hat{S}_H(V, \{\epsilon\}) = \left\{ z \in V;\, |z^{(0)}| < \epsilon_0 \right\} \bigcap \left(\bigcap_{p \in \Theta} p^{-1}(Y_H) \right).$$

Then $\hat{S}_H(V, \{\epsilon\})$ is a relatively compact Stein open subset and, by noticing Remark 4.1, the families $\{\hat{S}_H(V, \{\epsilon\})\}$ and $\{S_H(V, \{\epsilon\})\}$ of open sets are equivalent with respect to the inclusion of sets. Hence to complete the proof, it suffices to show the following fact: Let $U \subset \mathbb{C}^n$ be a subanalytic open subset. Assume U is relatively compact and Stein. Then we have

$$H^k(U; \mathcal{O}_X) = H^k(U; \mathcal{O}^t_X) = 0 \qquad (k \neq 0).$$

The fact for the sheaf $\mathcal{O}_X$ clearly holds by Oka's famous vanishing theorem (see Corollary 7.4.2 [13] for example), and the one for $\mathcal{O}^t_X$ is also well-known, see Theorem 2.10 [23]. This completes the proof. □

6 Examples of multi-specializations

In order to understand the previous constructions, in this section we construct several examples of multi-specializations. We shall work in the local model, in particular we are going to consider examples in $\mathbb{C}^n$ starting from the matrices defining the actions and the associated multi-normal deformations.

6.1 Majima's asymptotics

Let us consider $X = \mathbb{C}^n$ with coordinates $z = (z_1, \dots, z_n)$. We are going to check how our previous constructions permit to obtain strongly asymptotically developable functions in the sense of Majima [18]. Let $\chi = \{Z_1, \dots, Z_n\}$, $Z_i = \{z_i = 0\}$, $Z = \bigcap_{i=1}^n Z_i$. Set $\widetilde{X} = \mathbb{C}^n \times \mathbb{R}^n$ and consider the action

$$\begin{aligned} \mu_\chi : \mathbb{C}^n \times \mathbb{R}^n &\to \mathbb{C}^n, \\ (z_1, \dots, z_n, t_1, \dots, t_n) &\mapsto (z_1 t_1, \dots, z_n t_n). \end{aligned}$$

The monomials φ_i, $i = 1, \dots, n$, defining the action are $\varphi_i(t) = t_i$, $i = 1, \dots, n$, and the associated matrix A_χ is nothing but the identity matrix. By Proposition 1.10 the zero section S_χ of $\widetilde{X}$ has a vector bundle structure

$$S_\chi \simeq T_{Z_1} X \underset{X}{\times} \cdots \underset{X}{\times} T_{Z_n} X.$$

Following the notations of Section 2 we are going to study the family of multi-cones $C(\xi)$ associated with a point $\xi = (\xi_1, \dots, \xi_n) \in S_\chi$, $\xi_i \neq 0$, $i = 1, \dots, n$. First of all we easily see that the rational monomials φ_j^{-1}, $j = 1, \dots, n$ are $\varphi_j^{-1}(\tau) = \tau_j$ (no monomials ψ_k appear since A_χ is an invertible square matrix), and hence $F^0 = \{\tau_1, \dots, \tau_n\}$. A cofinal family of $C(\xi)$ is $C(\xi, F^0)$ defined as follows

$$\left\{ (z_1, \dots, z_n) \in X; \begin{matrix} z_i \in W_i & (k = 1, \dots, n), \\ |z_i| < \epsilon & \end{matrix} \right\},$$

where $\epsilon > 0$ and each W_i is an $\mathbb{R}^+$-conic open subset in $T_{Z_i}X \simeq \mathbb{C}$ containing the point ξ_i. It means that a cofinal family of $C(\xi)$ is given by the family of multi-sectors in $\mathbb{C}^n$ containing the direction ξ.

We are now ready to compute the stalks of the specialization of Whitney holomorphic functions. Let $\mathcal{O}_X^{\mathrm{w}}$ be the subanalytic sheaf of Whitney holomorphic functions on X. Let $\xi \in S_\chi \simeq \mathbb{C}^n$. Then

$$H^0(\rho^{-1}\nu_\chi^{sa}\mathcal{O}_X^{\mathrm{w}})_\xi \simeq \varinjlim_S H^0(S; \mathcal{O}_X^{\mathrm{w}}) \simeq \varinjlim_S \varprojlim_{S'} H^0(S'; \mathcal{O}_X^{\mathrm{w}}),$$

where $S = S_1 \times \cdots \times S_n$ ranges through the family of multi-sectors containing ξ and $S' = S_1' \times \cdots \times S_n'$ ranges through the family of multi-sectors properly contained in S (i.e. $\overline{S_i'} \setminus \{0\} \subset S_i$, $i = 1, \dots, n$). By Theorem 3.2 of [7],

$$\varprojlim_{S'} H^0(S'; \mathcal{O}_X^{\mathrm{w}}) \simeq \Gamma(S; \mathcal{A}_\chi),$$

where $\mathcal{A}_\chi$ denotes the sheaf of strongly asymptotically developable functions in the sense of Majima [18]. Moreover $H^k(\rho^{-1}\nu_\chi^{sa}(\mathcal{O}_X^{\mathrm{w}})) = 0$, $k \neq 0$, in view of Theorem 5.5, hence we get the isomorphism of sheaves (outside the fixed points of S_χ)

$$\rho^{-1}\nu_\chi^{sa}\mathcal{O}_X^{\mathrm{w}} \simeq \mathcal{A}_\chi.$$

Moreover, following the notations of Section 3, by Theorem 5.9 the sequence

$$0 \to \rho^{-1}\nu_\chi^{sa}\mathcal{O}^{\mathrm{w}}_{X|X\setminus\bigcup_k Z} \to \rho^{-1}\nu_\chi^{sa}\mathcal{O}_X^{\mathrm{w}} \to \rho^{-1}\nu_\chi^{sa}\mathcal{O}^{\mathrm{w}}_{X|\bigcup_k Z} \to 0$$

is exact. This corresponds to the sequence

$$0 \to \mathcal{A}_\chi^{<0} \to \mathcal{A}_\chi \to \mathcal{A}_\chi^{CF} \to 0,$$

where $\mathcal{A}_\chi^{<0}$ and $\mathcal{A}_\chi^{CF}$ denote the sheaves of flat asymptotics and consistent families of coefficients respectively. This is nothing but the Borel-Ritt exact sequence for strongly asymptotically developable holomorphic functions.

6.2 Examples in $\mathbb{C}^2$

Let us consider some interesting cases in $\mathbb{C}^2$ with variables $z = (z_1, z_2)$. We are going to consider the normal deformation (and the associated multi-specialization) for some interesting cases.

✓ **Majima specialization.** In this case

$$A_\chi = \begin{pmatrix} 1 & 0 \\ 0 & 1 \end{pmatrix}.$$

We have $Z_i = \{z_i = 0\}$, $i = 1, 2$. Then we can define a normal deformation $\widetilde{X} = \mathbb{C}^2 \times \mathbb{R}^2$ with the map $p(z; t) = (t_1 z_1, t_2 z_2)$. By Proposition 1.10 the zero section S_χ of $\widetilde{X}$ has a vector bundle structure

$$S_\chi \simeq T_{Z_1} X \underset{X}{\times} T_{Z_2} X.$$

The rational monomials φ_j^{-1}, $j = 1, 2$, are $\varphi_1^{-1}(\tau) = \tau_1$, $\varphi_2^{-1}(\tau) = \tau_2$. Let $\xi = (\xi_1, \xi_2)$, $\xi_i \neq 0$, $i = 1, 2$. A cofinal family of $C(\xi)$ is $C(\xi, F^0)$ consisting of elements $S(W, \epsilon)$, $W = W_1 \times W_2$, defined as follows

$$S(W, \epsilon) = \left\{ z \in X; \begin{matrix} z_k \in W_k & (k = 1, 2), \\ |z_i| < \epsilon & (i = 1, 2) \end{matrix} \right\},$$

where $\epsilon > 0$ and each W_k is an $\mathbb{R}^+$-conic open subset in $\mathbb{C}$ containing the point ξ_k.

✓ **Takeuchi specialization.** In this case

$$A_\chi = \begin{pmatrix} 1 & 1 \\ 0 & 1 \end{pmatrix}.$$

We have $Z_1 = \{0\}$, $Z_2 = \{z_2 = 0\}$. Then we can define a normal deformation $\widetilde{X} = \mathbb{C}^2 \times \mathbb{R}^2$ with the map $p(z; t) = (t_1 z_1, t_1 t_2 z_2)$. By Proposition 1.10 the zero section S_χ of $\widetilde{X}$ has a vector bundle structure

$$S_\chi \simeq T_{Z_1} Z_2 \underset{X}{\times} T_{Z_2} X.$$

The rational monomials φ_j^{-1}, $j = 1, 2$, are $\varphi_1^{-1}(\tau) = \tau_1$, $\varphi_2^{-1}(\tau) = \dfrac{\tau_2}{\tau_1}$. Let $\xi = (\xi_1, \xi_2)$, $\xi_i \neq 0$, $i = 1, 2$. A cofinal family of $C(\xi)$ is $C(\xi, F^0)$ consisting of elements $S(W, \epsilon)$, $W = W_1 \times W_2$, defined as follows

$$S(W, \epsilon) = \left\{ \begin{matrix} z_k \in W_k \quad (k = 1, 2), \\ z \in X; |z_1| < \epsilon, \\ |z_2| < \epsilon |z_1| \end{matrix} \right\},$$

where $\epsilon > 0$ and each W_k is an $\mathbb{R}^+$-conic open subset in $\mathbb{C}$ containing the point ξ_k. This is the bispecialization of [24].

Set $X = Y = \mathbb{C}^2$, X (resp. Y) with coordinates $z = (z_1, z_2)$ (resp. $w = (w_1, w_2)$). Let $\chi_Y = \{Z_{Y1}, Z_{Y2}\}$, with $Z_{Y1} = \{0\}$, $Z_{Y2} = \{w_2 = 0\}$ and $\chi_X = \{Z_{X1}, Z_{X2}\}$, with $Z_{X1} = f^{-1}(0) = \{z_1 = 0\}$, $Z_{X2} = \{z_2 = 0\}$. We have

$$A_{\chi_X} = \begin{pmatrix} 1 & 0 \\ 0 & 1 \end{pmatrix}, \quad A_{\chi_Y} = \begin{pmatrix} 1 & 1 \\ 0 & 1 \end{pmatrix}.$$

Consider the map

$$\begin{aligned} f : X &\to Y \\ (z_1, z_2) &\mapsto (z_1, z_1 z_2) \end{aligned}$$

(locally) the blow-up at the origin. Remark that condition (3.3) holds. Let us consider $\widetilde{f}$. On the zero section (we keep the same coordinates for simplicity) the map $T_\chi f$ is defined as follows:

$$\begin{aligned} w_1 &= \frac{\partial f}{\partial z_1}(0,0) z_1 = z_1, \\ w_2 &= \frac{\partial^2 f}{\partial z_1 \partial z_2}(0,0) z_1 z_2 = z_1 z_2. \end{aligned}$$

It is a conic map with respect to the $(\mathbb{R}^+)^2$-actions on S_{χ_X} and S_{χ_Y}. By Propositions 3.18 and 3.19, f makes a link between specializations with respect to χ_X and χ_Y.

✓ **Cusp specialization.** In this case

$$A_\chi = \begin{pmatrix} 3 & 2 \\ 1 & 1 \end{pmatrix}.$$

We have $Z_1 = Z_2 = \{0\}$. Then we can define a normal deformation $\widetilde{X} = \mathbb{C}^2 \times \mathbb{R}^2$ with the map $p(z; t) = (t_1^3 t_2 z_1, t_1^2 t_2 z_2)$. The rational monomials φ_j^{-1}, $j = 1, 2$, are $\varphi_1^{-1}(\tau) = \frac{\tau_1}{\tau_2}$, $\varphi_2^{-1}(\tau) = \frac{\tau_2^3}{\tau_1^2}$. Let $\xi = (\xi_1, \xi_2)$, $\xi_i \neq 0$, $i = 1, 2$. A cofinal family of $C(\xi)$ is $C(\xi, F^0)$ consisting of elements $S(W, \epsilon)$, $W = W_1 \times W_2$, defined as follows

$$S(W, \epsilon) = \left\{ z \in X; \begin{array}{c} z_k \in W_k \quad (k = 1, 2), \\ |z_1| < \epsilon |z_2|, \\ |z_2|^3 < \epsilon |z_1|^2 \end{array} \right\},$$

where $\epsilon > 0$ and each W_k is an $\mathbb{R}^+$-conic open subset in $\mathbb{C}$ containing the point ξ_k.

Set $X = Y = \mathbb{C}^2$, X (resp. Y) with coordinates $z = (z_1, z_2)$ (resp. $w = (w_1, w_2)$). Let $\chi_Y = \{Z_{Y1}, Z_{Y2}\}$, with $Z_{Y1} = Z_{Y2} = \{0\}$ and $\chi_X = \{Z_{X1}, Z_{X2}\}$, with $Z_{X1} = \{z_1 = 0\}$, $Z_{X2} = \{z_2 = 0\}$. We have

$$A_{\chi_X} = \begin{pmatrix} 1 & 0 \\ 0 & 1 \end{pmatrix}, \quad A_{\chi_Y} = \begin{pmatrix} 3 & 2 \\ 1 & 1 \end{pmatrix}.$$

Consider the map

$$\begin{aligned} f : X &\to Y \\ (z_1, z_2) &\mapsto (z_1^3 z_2, z_1^2 z_2) \end{aligned}$$

(desingularization of a cusp at the origin). Remark that condition (3.3) holds. Let us consider $\widetilde{f}$. On the zero section (we keep the same coordinates for simplicity) the map $T_\chi f$ is defined as follows:

$$\begin{aligned} w_1 &= \frac{1}{4!} \cdot 4 \cdot \frac{\partial^4 f_1}{\partial z_1^3 \partial z_2}(0,0) z_1^3 z_2 = z_1^3 z_2, \\ w_2 &= \frac{1}{3!} \cdot 3 \cdot \frac{\partial^3 f_1}{\partial z_1^2 \partial z_2}(0,0) z_1^2 z_2 = z_1^2 z_2. \end{aligned}$$

It is a conic map with respect to the $(\mathbb{R}^+)^2$-actions on S_{χ_X} and S_{χ_Y}. By Propositions 3.18 and 3.19, f makes a link between specializations with respect to χ_X and χ_Y.

6.3 Examples in $\mathbb{C}^3$

Let us consider some interesting cases in $\mathbb{C}^3$ with variables $z = (z_1, z_2, z_3)$. We are going to consider the normal deformation (and the associated multi-specialization) for some interesting cases. We shall consider 2×3 and 3×3 matrices with entries $0, 1$, up to permutation of coordinates, avoiding the degenerate cases.

✓ **Clean intersection specialization (2 lines).** In this case

$$A_\chi = \begin{pmatrix} 1 & 1 & 0 \\ 0 & 1 & 1 \end{pmatrix}.$$

We have $Z_1 = \{z_1 = z_2 = 0\}$, $Z_2 = \{z_2 = z_3 = 0\}$. Then we can define a normal deformation $\widetilde{X} = \mathbb{C}^3 \times \mathbb{R}^2$ with the map $p(z; t) = (t_1 z_1, t_1 t_2 z_2, t_2 z_3)$. The rational monomials φ_j^{-1}, $j = 1, 2$, are $\varphi_1^{-1}(\tau) = \tau_1$, $\varphi_2^{-1}(\tau) = \dfrac{\tau_2}{\tau_1}$ and $\psi_3(\tau) = \dfrac{\tau_1 \tau_3}{\tau_2}$. Let $\xi = (\xi_1, \xi_2, \xi_3)$, $\xi_i \neq 0$, $i = 1, 2$. A cofinal family of $C(\xi)$ is $C(\xi, F^0)$ consisting of elements $S(W,\epsilon)$, $W = W_1 \times W_2 \times W_3$, defined as follows:

$$S(W,\epsilon) = \left\{ z \in X; \begin{array}{l} z_k \in W_k \quad (k = 1,2,3), \\ |z_1| < \epsilon, \\ |z_2| < \epsilon |z_1|, \\ (n_3(\xi) - \epsilon)|z_2| < |z_1||z_3| < (n_3(\xi) + \epsilon)|z_2| \end{array} \right\},$$

where $n_3(\xi) := |\xi_1 \xi_3 / \xi_2|$, $\epsilon > 0$ and each W_k is an $\mathbb{R}^+$-conic open subset in $\mathbb{C}$ containing the point ξ_k.

✓ **Majima specialization.** In this case

$$A_\chi = \begin{pmatrix} 1 & 0 & 0 \\ 0 & 1 & 0 \\ 0 & 0 & 1 \end{pmatrix}.$$

We have $Z_i = \{z_i = 0\}$, $i = 1,2,3$. Then we can define a normal deformation $\widetilde{X} = \mathbb{C}^3 \times \mathbb{R}^3$ with the map $p(z;\, t) = (t_1 z_1, t_2 z_2, t_3 z_3)$. By Proposition 1.10 the zero section S_χ of $\widetilde{X}$ has a vector bundle structure

$$S_\chi \simeq T_{Z_1} X \underset{X}{\times} T_{Z_2} X \underset{X}{\times} T_{Z_3} X.$$

The rational monomials φ_j^{-1}, $j = 1,2,3$, are $\varphi_1^{-1}(\tau) = \tau_1$, $\varphi_2^{-1}(\tau) = \tau_2$, $\varphi_3^{-1}(\tau) = \tau_3$. Let $\xi = (\xi_1, \xi_2, \xi_3)$, $\xi_i \neq 0$, $i = 1,2,3$. A cofinal family of $C(\xi)$ is $C(\xi, F^0)$ consisting of elements $S(W,\epsilon)$, $W = W_1 \times W_2 \times W_3$, defined as follows

$$S(W,\epsilon) = \left\{ z \in X; \begin{array}{ll} z_k \in W_k & (k = 1,2,3), \\ |z_i| < \epsilon & (i = 1,2,3) \end{array} \right\},$$

where $\epsilon > 0$ and each W_k is an $\mathbb{R}^+$-conic open subset in $\mathbb{C}$ containing the point ξ_k. This is the multi-specialization of [9] (Majima case) in $\mathbb{C}^3$.

✓ **Takeuchi specialization.** In this case

$$A_\chi = \begin{pmatrix} 1 & 1 & 1 \\ 0 & 1 & 1 \\ 0 & 0 & 1 \end{pmatrix}.$$

We have $Z_1 = \{0\}$, $Z_2 = \{z_2 = z_3 = 0\}$, $Z_3 = \{z_3 = 0\}$. Then we can define a normal deformation $\widetilde{X} = \mathbb{C}^3 \times \mathbb{R}^3$ with the map $p(z;\, t) = (t_1 z_1, t_1 t_2 z_2, t_1 t_2 t_3 z_3)$. By Proposition 1.10 the zero section S_χ of $\widetilde{X}$ has a vector bundle structure

$$S_\chi \simeq T_{Z_1} Z_2 \underset{X}{\times} T_{Z_2} Z_3 \underset{X}{\times} T_{Z_3} X.$$

The rational monomials φ_j^{-1}, $j=1,2,3$, are $\varphi_1^{-1}(\tau)=\tau_1$, $\varphi_2^{-1}(\tau)=\dfrac{\tau_2}{\tau_1}$, $\varphi_3^{-1}(\tau)=\dfrac{\tau_3}{\tau_2}$. Let $\xi=(\xi_1,\xi_2,\xi_3)$, $\xi_i\neq 0$, $i=1,2,3$. A cofinal family of $C(\xi)$ is $C(\xi,F^0)$ consisting of elements $S(W,\epsilon)$, $W=W_1\times W_2\times W_3$, defined as follows

$$S(W,\epsilon)=\left\{z\in X;\begin{array}{l} z_k\in W_k \quad (k=1,2,3),\\ |z_1|<\epsilon,\\ |z_2|<\epsilon|z_1|,\\ |z_3|<\epsilon|z_2|\end{array}\right\},$$

where $\epsilon>0$ and each W_k is an $\mathbb{R}^+$-conic open subset in $\mathbb{C}$ containing the point ξ_k. This is the multi-specialization of [9] (Takeuchi case) in $\mathbb{C}^3$.

✓ **2 blocks (Majima-Takeuchi).** In this case

$$A_\chi=\begin{pmatrix}1&0&0\\0&1&1\\0&0&1\end{pmatrix}.$$

We have $Z_1=\{z_1=0\}$, $Z_2=\{z_2=z_3=0\}$, $Z_3=\{z_3=0\}$. Then we can define a normal deformation $\widetilde{X}=\mathbb{C}^3\times\mathbb{R}^3$ with the map $p(z;t)=(t_1z_1,t_2z_2,t_2t_3z_3)$. By Proposition 1.10 the zero section S_χ of $\widetilde{X}$ has a vector bundle structure

$$S_\chi\simeq T_{Z_1}X\underset{X}{\times}T_{Z_2}Z_3\underset{X}{\times}T_{Z_3}X.$$

The rational monomials φ_j^{-1}, $j=1,2,3$, are $\varphi_1^{-1}(\tau)=\tau_1$, $\varphi_2^{-1}(\tau)=\tau_2$, $\varphi_3^{-1}(\tau)=\dfrac{\tau_3}{\tau_2}$. Let $\xi=(\xi_1,\xi_2,\xi_3)$, $\xi_i\neq 0$, $i=1,2,3$. A cofinal family of $C(\xi)$ is $C(\xi,F^0)$ consisting of elements $S(W,\epsilon)$, $W=W_1\times W_2\times W_3$, defined as follows:

$$S(W,\epsilon)=\left\{z\in X;\begin{array}{l} z_k\in W_k \quad (k=1,2,3),\\ |z_1|<\epsilon,\\ |z_2|<\epsilon,\\ |z_3|<\epsilon|z_2|\end{array}\right\},$$

where $\epsilon>0$ and each W_k is an $\mathbb{R}^+$-conic open subset in $\mathbb{C}$ containing the point ξ_k.

✓ **Clean intersection (3 lines) specialization.** In this case

$$A_\chi=\begin{pmatrix}0&1&1\\1&0&1\\1&1&0\end{pmatrix}.$$

We have $Z_1 = \{z_2 = z_3 = 0\}$, $Z_2 = \{z_1 = z_3 = 0\}$, $Z_3 = \{z_1 = z_2 = 0\}$. Then we can define a normal deformation $\widetilde{X} = \mathbb{C}^3 \times \mathbb{R}^3$ with the map $p(z;\, t) = (t_2t_3z_1, t_1t_3z_2, t_1t_2z_3)$. Set $Z = Z_1 \cap Z_2 \cap Z_3$. By Proposition 1.10 the zero section S_χ of $\widetilde{X}$ has a vector bundle structure

$$S_\chi \simeq T_Z Z_1 \underset{X}{\times} T_Z Z_2 \underset{X}{\times} T_Z Z_3.$$

The rational monomials φ_j^{-1}, $j = 1, 2, 3$, are $\varphi_1^{-1}(\tau) = \left(\dfrac{\tau_2\tau_3}{\tau_1}\right)^{\frac{1}{2}}$, $\varphi_2^{-1}(\tau) = \left(\dfrac{\tau_1\tau_3}{\tau_2}\right)^{\frac{1}{2}}$, $\varphi_3^{-1}(\tau) = \left(\dfrac{\tau_1\tau_2}{\tau_3}\right)^{\frac{1}{2}}$. Let $\xi = (\xi_1, \xi_2, \xi_3)$, $\xi_i \neq 0$, $i = 1, 2, 3$. A cofinal family of $C(\xi)$ is $C(\xi, F^0)$ consisting of elements $S(W, \epsilon)$, $W = W_1 \times W_2 \times W_3$, defined as follows

$$S(W,\epsilon) = \left\{ z \in X;\; \begin{array}{l} z_k \in W_k \quad (k = 1,2,3), \\ |z_2||z_3| < \epsilon |z_1|, \\ |z_1||z_3| < \epsilon |z_2|, \\ |z_1||z_2| < \epsilon |z_3| \end{array} \right\},$$

where $\epsilon > 0$ and each W_k is an $\mathbb{R}^+$-conic open subset in $\mathbb{C}$ containing the point ξ_k.

✓ **Mixed specialization (Majima-Takeuchi).** In this case

$$A_\chi = \begin{pmatrix} 1 & 1 & 1 \\ 0 & 1 & 0 \\ 0 & 0 & 1 \end{pmatrix}.$$

We have $Z_1 = \{0\}$, $Z_2 = \{z_2 = 0\}$, $Z_3 = \{z_3 = 0\}$, Then we can define a normal deformation $\widetilde{X} = \mathbb{C}^3 \times \mathbb{R}^3$ with the map $p(z;\, t) = (t_1z_1, t_1t_2z_2, t_1t_3z_3)$. By Proposition 1.10 the zero section S_χ of $\widetilde{X}$ has a vector bundle structure

$$S_\chi \simeq T_{Z_1}(Z_2 \cap Z_3) \underset{X}{\times} T_{Z_2}X \underset{X}{\times} T_{Z_3}X.$$

The rational monomials φ_j^{-1}, $j = 1, 2, 3$, are $\varphi_1^{-1}(\tau) = \tau_1$, $\varphi_2^{-1}(\tau) = \dfrac{\tau_2}{\tau_1}$, $\varphi_3^{-1}(\tau) = \dfrac{\tau_3}{\tau_1}$. Let $\xi = (\xi_1, \xi_2, \xi_3)$, $\xi_i \neq 0$, $i = 1, 2, 3$. A cofinal family of $C(\xi)$ is $C(\xi, F^0)$ consisting of elements $S(W, \epsilon)$, $W = W_1 \times W_2 \times W_3$, defined as follows

$$S(W,\epsilon) = \left\{ z \in X;\; \begin{array}{l} z_k \in W_k \quad (k = 1,2,3), \\ |z_1| < \epsilon, \\ |z_2| < \epsilon |z_1|, \\ |z_3| < \epsilon |z_1| \end{array} \right\},$$

where $\epsilon > 0$ and each W_k is an $\mathbb{R}^+$-conic open subset in $\mathbb{C}$ containing the point ξ_k. This is the multi-specialization of [9] (mixed case) in $\mathbb{C}^3$.

✓ **Mixed specialization (2 lines clean intersection-Takeuchi I).** In this case

$$A_\chi = \begin{pmatrix} 1 & 1 & 0 \\ 0 & 1 & 1 \\ 0 & 0 & 1 \end{pmatrix}.$$

We have $Z_1 = \{z_1 = z_2 = 0\}$, $Z_2 = \{z_2 = z_3 = 0\}$, $Z_3 = \{z_3 = 0\}$. Then we can define a normal deformation $\widetilde{X} = \mathbb{C}^3 \times \mathbb{R}^3$ with the map $p(z;\, t) = (t_1 z_1, t_1 t_2 z_2, t_2 t_3 z_3)$. Set $Z = Z_1 \cap Z_2$. By Proposition 1.10 the zero section S_χ of $\widetilde{X}$ has a vector bundle structure

$$S_\chi \simeq T_Z Z_2 \underset{X}{\times} \frac{TX \underset{X}{\times} Z}{TZ_1 \underset{X}{\times} Z + TZ_2 \underset{X}{\times} Z} \underset{X}{\times} T_X Z_3.$$

The rational monomials φ_j^{-1}, $j = 1, 2, 3$, are $\varphi_1^{-1}(\tau) = \tau_1$, $\varphi_2^{-1}(\tau) = \dfrac{\tau_2}{\tau_1}$, $\varphi_3^{-1}(\tau) = \dfrac{\tau_1\tau_3}{\tau_2}$. Let $\xi = (\xi_1, \xi_2, \xi_3)$, $\xi_i \neq 0$, $i = 1, 2, 3$. A cofinal family of $C(\xi)$ is $C(\xi, F^0)$ consisting of elements $S(W, \epsilon)$, $W = W_1 \times W_2 \times W_3$, defined as follows:

$$S(W,\epsilon) = \left\{ z \in X; \begin{array}{l} z_k \in W_k \quad (k = 1, 2, 3), \\ |z_1| < \epsilon, \\ |z_2| < \epsilon |z_1|, \\ |z_1||z_3| < \epsilon |z_2| \end{array} \right\},$$

where $\epsilon > 0$ and each W_k is an $\mathbb{R}^+$-conic open subset in $\mathbb{C}$ containing the point ξ_k.

✓ **Mixed specialization (2 lines clean intersection-Takeuchi II).** In this case

$$A_\chi = \begin{pmatrix} 0 & 1 & 1 \\ 1 & 0 & 1 \\ 0 & 0 & 1 \end{pmatrix}.$$

We have $Z_1 = \{z_2 = z_3 = 0\}$, $Z_2 = \{z_1 = z_3 = 0\}$, $Z_3 = \{z_3 = 0\}$. Then we can define a normal deformation $\widetilde{X} = \mathbb{C}^3 \times \mathbb{R}^3$ with the map $p(z;\, t) = (t_2 z_1, t_1 z_2, t_1 t_2 t_3 z_3)$. Set $Z = Z_1 \cap Z_2$. By Proposition 1.10 the zero section S_χ of $\widetilde{X}$ has a vector bundle structure

$$S_\chi \simeq T_Z Z_1 \underset{X}{\times} T_Z Z_2 \underset{X}{\times} T_{Z_3} X.$$

The rational monomials φ_j^{-1}, $j = 1, 2, 3$, are $\varphi_1^{-1}(\tau) = \tau_2$, $\varphi_2^{-1}(\tau) = \tau_1$, $\varphi_3^{-1}(\tau) = \dfrac{\tau_3}{\tau_1\tau_2}$. Let $\xi = (\xi_1, \xi_2, \xi_3)$, $\xi_i \neq 0$, $i = 1, 2, 3$. A cofinal family of $C(\xi)$ is $C(\xi, F^0)$

consisting of elements $S(W,\epsilon)$, $W = W_1 \times W_2 \times W_3$, defined as follows:

$$S(W,\epsilon) = \left\{ z \in X; \begin{array}{l} z_k \in W_k \quad (k=1,2,3), \\ |z_1| < \epsilon, \\ |z_2| < \epsilon, \\ |z_3| < \epsilon |z_1||z_2| \end{array} \right\},$$

where $\epsilon > 0$ and each W_k is an $\mathbb{R}^+$-conic open subset in $\mathbb{C}$ containing the point ξ_k.

✓ **Mixed specialization (2 lines clean intersection-Takeuchi III).** In this case

$$A_\chi = \begin{pmatrix} 1 & 1 & 1 \\ 0 & 1 & 1 \\ 1 & 0 & 1 \end{pmatrix}.$$

We have $Z_1 = \{0\}$, $Z_2 = \{z_2 = z_3 = 0\}$, $Z_3 = \{z_1 = z_3 = 0\}$. Then we can define a normal deformation $\widetilde{X} = \mathbb{C}^3 \times \mathbb{R}^3$ with the map $p(z;t) = (t_1t_3z_1, t_1t_2z_2, t_1t_2t_3z_3)$. By Proposition 1.10 the zero section S_χ of $\widetilde{X}$ has a vector bundle structure

$$S_\chi \simeq T_{Z_1}Z_2 \underset{X}{\times} T_{Z_1}Z_3 \underset{X}{\times} \frac{TX \underset{X}{\times} Z_1}{TZ_2 \underset{X}{\times} Z_1 + TZ_3 \underset{X}{\times} Z_1}.$$

The rational monomials φ_j^{-1}, $j=1,2,3$, are $\varphi_1^{-1}(\tau) = \dfrac{\tau_1\tau_2}{\tau_3}$, $\varphi_2^{-1}(\tau) = \dfrac{\tau_3}{\tau_1}$, $\varphi_3^{-1}(\tau) = \dfrac{\tau_3}{\tau_2}$. Let $\xi = (\xi_1, \xi_2, \xi_3)$, $\xi_i \neq 0$, $i = 1,2,3$. A cofinal family of $C(\xi)$ is $C(\xi, F^0)$ consisting of elements $S(W,\epsilon)$, $W = W_1 \times W_2 \times W_3$, defined as follows:

$$S(W,\epsilon) = \left\{ z \in X; \begin{array}{l} z_k \in W_k \quad (k=1,2,3), \\ |z_1||z_2| < \epsilon |z_3|, \\ |z_3| < \epsilon |z_1|, \\ |z_3| < \epsilon |z_2| \end{array} \right\},$$

where $\epsilon > 0$ and each W_k is an $\mathbb{R}^+$-conic open subset in $\mathbb{C}$ containing the point ξ_k.

6.4 Classification of multi-specialization for the case of two submanifolds

Let us now consider $X = \mathbb{C}^n$ with variables $z = (z^{(1)}, \ldots, z^{(m)})$, where $z^{(k)}$, $k = 1, \ldots, m$, denote the coordinates blocks $(z_{i_{k,1}}, \ldots, z_{i_{k,n_k}})$ (we forget $z^{(0)}$ to lighten notations). We consider the normal deformation (and the associated multi-specialization) in the two manifold case, i.e., $\chi = \{Z_1, Z_2\}$ with $Z_j = \{z^{(k)} = 0,\ k \in K_j\}$, $K_j \subseteq \{1, \ldots, m\}$, $j = 1,2$. Remark that Z_1 and Z_2 may coincide.

The action associated with χ is now defined by

$$\mu_j(z,\lambda) = (\lambda^{a_{j1}} z^{(1)}, \ldots, \lambda^{a_{jm}} z^{(m)}) \qquad (j = 1,2)$$

with a_{ji} non-negative rationals, $a_{ji} \neq 0$ if $i \in K_j$, $a_{ji} = 0$ otherwise. The associated matrix A_χ is

$$\begin{pmatrix} a_{11} & \cdots & a_{1m} \\ a_{21} & \cdots & a_{2m} \end{pmatrix}$$

and we assume for simplicity that all the columns are non zero. We also assume that the action is non-degenerate and that the matrix A'_χ consisting of the first two columns of A_χ is invertible. Then we can define a normal deformation $\widetilde{X} = \mathbb{C}^n \times \mathbb{R}^2$ with the map $p : \widetilde{X} \to X$ defined by

$$p(z;\, t) = (z^{(0)},\, \varphi_1(t) z^{(1)}, \ldots, \varphi_m(t) z^{(m)})$$

with

$$\varphi_k(t) = t_1^{a_{1k}} t_2^{a_{2k}} \qquad (k = 1, 2, \ldots, m). \tag{6.1}$$

Following the notations of Section 2 we are going to study the family of multi-cones $C(\xi)$ associated with a point $\xi = (\xi_1, \ldots, \xi_n) \in S_\chi$. The rational monomials φ_j^{-1}, $j = 1, 2$ are determined by A'^{-1}_χ, where

$$A'_\chi = \begin{pmatrix} a_{11} & a_{12} \\ a_{21} & a_{22} \end{pmatrix} \quad \text{and} \quad A'^{-1}_\chi = \frac{1}{a_{11}a_{22} - a_{12}a_{21}} \begin{pmatrix} a_{22} & -a_{12} \\ -a_{21} & a_{11} \end{pmatrix}.$$

Setting $d = \dfrac{1}{a_{11}a_{22} - a_{12}a_{21}}$, we have

$$\varphi_1^{-1}(\tau) = (\tau_1^{a_{22}} \tau_2^{-a_{21}})^d, \quad \varphi_2^{-1}(\tau) = (\tau_1^{-a_{12}} \tau_2^{a_{11}})^d.$$

For $k \geq 3$ we have

$$\psi_k(\tau) = \frac{\tau_k}{(\tau_1^{a_{1k}a_{22} - a_{2k}a_{12}} \tau_2^{a_{2k}a_{11} - a_{1k}a_{21}})^d}.$$

Here we may assume $d > 0$ by exchanging the first and the second columns of A_χ if necessary. Suppose that ξ is outside of the set of fixed points. By definition (up to take a permutation of coordinates blocks), we may assume that $\xi^{(1)}, \xi^{(2)} \neq 0$. Further, for simplicity, we also assume ξ to be normalized, that is, $|\xi^{(1)}| = |\xi^{(2)}| = 1$. A cofinal family of $C(\xi)$ is $C(\xi, F^0)$ consisting of elements $S(W,\epsilon)$, $W = W_1 \times \cdots \times W_m$, defined as follows

$$S(W,\epsilon) = \left\{ z \in X;\ \begin{array}{l} z^{(k)} \in W_k \quad (k = 1, \ldots, m), \\ |z^{(i)}|^{a_{jj}} < \epsilon^{\frac{1}{d}} |z^{(j)}|^{a_{ji}} \quad (i, j \in \{1,2\},\ i \neq j), \\ |\xi^{(k)}| - \epsilon < \dfrac{|z^{(k)}|}{(|z^{(1)}|^{a_{1k}a_{22} - a_{2k}a_{12}} |z^{(2)}|^{a_{2k}a_{11} - a_{1k}a_{21}})^d} \\ \qquad < |\xi^{(k)}| + \epsilon \quad (k \geq 3) \end{array} \right\},$$

where $\epsilon > 0$ and each W_k is an $\mathbb{R}^+$-conic open subset in $\mathbb{C}^{n_k}$ containing the point $\xi^{(k)}$.

We are now ready to compute the stalks of the specialization of Whitney holomorphic functions at ξ. Let $\mathcal{O}_X^{\mathrm{w}}$ be the subanalytic sheaf of Whitney holomorphic functions on X. Then

$$H^0(\nu_\chi \mathcal{O}_X^{\mathrm{w}})_\xi \simeq \varinjlim_S H^0(S; \mathcal{O}_X^{\mathrm{w}}) \simeq \varinjlim_S \varprojlim_{S'} H^0(S'; \mathcal{O}_X^{\mathrm{w}}),$$

where S ranges through the family $\{S(W, \epsilon)\}_{W,\epsilon}$, S' ranges through the family of multi-cones properly contained in S.

We are now going to classify multi-specializations in the two manifolds case for $m \leq 3$. We perform the classification by

- the number m of columns of A_χ,
- the number of non-zero entries of A_χ (denoted by N) assuming that the action is non-degenerate (in that case we are reduced to the one manifold case)

If $m = 1$ the action is degenerate, so we consider $m = 2, 3$. Moreover, the fact that row (resp. column) permutation and row (resp. column) multiplication by a positive rational number do not change the multi-normal deformation (up to permutation of variable blocks) , we may classify A_χ up to these operations. In particular, we assume $a_{11} = a_{22} = 1$ and $1 - a_{12}a_{21} > 0$.

Let us consider the case $m = 2$. We have 2 coordinates blocks $(m_1 + m_2 = n)$

$$z^{(1)} = \{z_1, \ldots, z_{m_1}\}, \quad z^{(2)} = \{z_{m_1+1}, \ldots, z_{m_1+m_2}\}.$$

In this case there are 5 possibilities.

✓ **Case $m = 2$ and $N = 0, 1$.** The action is degenerate.

✓ **Case $m = 2, N = 2$.** In this case

$$A_\chi = \begin{pmatrix} 1 & 0 \\ 0 & 1 \end{pmatrix}.$$

We have $Z_j = \{z^{(j)} = 0\}$, $j = 1, 2$. Then we can define a normal deformation $\widetilde{X} = \mathbb{C}^n \times \mathbb{R}^2$ with the map $p(z; t) = (t_1 z^{(1)}, t_2 z^{(2)})$. The rational monomials φ_j^{-1}, $j = 1, 2$ are $\varphi_1^{-1}(\tau) = \tau_1$ and $\varphi_2^{-1}(\tau) = \tau_2$. A cofinal family of $C(\xi)$ is $C(\xi, F^0)$

consisting of elements $S(W,\epsilon)$, $W = W_1 \times W_2$, defined as follows

$$S(W,\epsilon) = \left\{ z \in X; \begin{array}{ll} z^{(k)} \in W_k & (k = 1,2), \\ |z^{(i)}| < \epsilon & (i = 1,2) \end{array} \right\},$$

where $\epsilon > 0$ and each W_k is an $\mathbb{R}^+$-conic open subset in $\mathbb{C}^{m_k}$ containing the point $\xi^{(k)}$.

✓ **Case $m = 2, N = 3$.** In this case

$$A_\chi = \begin{pmatrix} 1 & b \\ 0 & 1 \end{pmatrix},$$

with $b \in \mathbb{Q}_{>0}$. We have $Z_1 = \{0\}$, $Z_2 = \{z^{(2)} = 0\}$. Then we can define a normal deformation $\widetilde{X} = \mathbb{C}^n \times \mathbb{R}^2$ with the map $p(z;t) = (t_1 z^{(1)}, t_1^b t_2 z^{(2)})$. The rational monomials φ_j^{-1}, $j = 1,2$, are $\varphi_1^{-1}(\tau) = \tau_1$ and $\varphi_2^{-1}(\tau) = \dfrac{\tau_2}{\tau_1^b}$. A cofinal family of $C(\xi)$ is $C(\xi, F^0)$ consisting of elements $S(W,\epsilon)$, $W = W_1 \times W_2$, defined as follows

$$S(W,\epsilon) = \left\{ \begin{array}{l} \qquad\quad z^{(k)} \in W_k \quad (k = 1,2), \\ z \in X; |z^{(1)}| < \epsilon, \\ \qquad\quad |z^{(2)}| < \epsilon |z^{(1)}|^b \end{array} \right\},$$

where $\epsilon > 0$ and each W_k is an $\mathbb{R}^+$-conic open subset in $\mathbb{C}^{m_k}$ containing the point $\xi^{(k)}$.

✓ **Case $m = 2, N = 4$.** In this case

$$A_\chi = \begin{pmatrix} 1 & b \\ c & 1 \end{pmatrix},$$

with $b, c \in \mathbb{Q}_{>0}$ and $1 - bc > 0$. We have $Z_1 = Z_2 = \{0\}$. Then we can define a normal deformation $\widetilde{X} = \mathbb{C}^n \times \mathbb{R}^2$ with the map $p(z;t) = (t_1 t_2^c z^{(1)}, t_1^b t_2 z^{(2)})$. Set $d = \dfrac{1}{1 - bc}$. The rational monomials φ_j^{-1}, $j = 1,2$ are $\varphi_1^{-1}(\tau) = \left(\dfrac{\tau_1}{\tau_2^c}\right)^d$ and $\varphi_2^{-1}(\tau) = \left(\dfrac{\tau_2}{\tau_1^b}\right)^d$. A cofinal family of $C(\xi)$ is $C(\xi, F^0)$ consisting of elements $S(W,\epsilon)$, $W = W_1 \times W_2$, defined as follows

$$S(W,\epsilon) = \left\{ \begin{array}{l} \qquad\quad z^{(k)} \in W_k \quad (k = 1,2), \\ z \in X; |z^{(1)}| < \epsilon^{\frac{1}{d}} |z^{(2)}|^c, \\ \qquad\quad |z^{(2)}| < \epsilon^{\frac{1}{d}} |z^{(1)}|^b \end{array} \right\},$$

where $\epsilon > 0$ and each W_k is an $\mathbb{R}^+$-conic open subset in $\mathbb{C}^{m_k}$ containing the point $\xi^{(k)}$.

Let us consider the case $m = 3$. We have 3 coordinates blocks

$$\begin{aligned} z^{(1)} &= \{z_1, \ldots, z_{m_1}\}, \\ z^{(2)} &= \{z_{m_1+1}, \ldots, z_{m_1+m_2}\}, \\ z^{(3)} &= \{z_{m_1+m_2+1}, \ldots, z_{m_1+m_2+m_3}\}, \end{aligned}$$

with $m_1 + m_2 + m_3 = n$. Up to row (resp. column) permutation and row (resp. column) multiplication by a positive rational number there are 5 interesting cases.

✓ **Case $m = 3, N = 3$.** In this case

$$A_\chi = \begin{pmatrix} 1 & 0 & e \\ 0 & 1 & 0 \end{pmatrix}$$

with $1 \neq e \in \mathbb{Q}_{>0}$. We have $Z_1 = \{z^{(1)} = z^{(3)} = 0\}$ and $Z_2 = \{z^{(2)} = 0\}$. Then we can define a normal deformation $\widetilde{X} = \mathbb{C}^n \times \mathbb{R}^2$ with the map $p(z; t) = (t_1 z^{(1)}, t_2 z^{(2)}, t_1^e z^{(3)})$. The rational monomials φ_j^{-1}, $j = 1, 2$ are $\varphi_1^{-1}(\tau) = \tau_1$, $\varphi_2^{-1}(\tau) = \tau_2$ and $\psi_3 = \dfrac{\tau_3}{\tau_1^e}$. A cofinal family of $C(\xi)$ is $C(\xi, F^0)$ consisting of elements $S(W, \epsilon)$, $W = W_1 \times W_2 \times W_3$, defined as follows

$$S(W, \epsilon) = \left\{ z \in X; \begin{array}{c} z^{(k)} \in W_k \quad (k = 1, 2, 3), \\ |z^{(j)}| < \epsilon \quad (j = 1, 2), \\ (|\xi^{(3)}| - \epsilon)|z^{(1)}|^e < |z^{(3)}| < (|\xi^{(3)}| + \epsilon)|z^{(1)}|^e \end{array} \right\},$$

where $\epsilon > 0$ and each W_k is an $\mathbb{R}^+$-conic open subset in $\mathbb{C}^{m_k}$ containing the point $\xi^{(k)}$.

✓ **Case $m = 3, N = 4$.** In this case we have two possibilities

(a) $\quad A_\chi = \begin{pmatrix} 1 & b & e \\ 0 & 1 & 0 \end{pmatrix} \quad b, e \in \mathbb{Q}_{>0},\ e \neq 1,$

(b) $\quad A_\chi = \begin{pmatrix} 1 & b & 0 \\ 0 & 1 & e \end{pmatrix} \quad b, e \in \mathbb{Q}_{>0}.$

✓ **Case** $m=3, N=4$ **(a).** We have $Z_1=\{0\}$ and $Z_2=\{z^{(2)}=0\}$. Then we can define a normal deformation $\widetilde{X}=\mathbb{C}^n\times\mathbb{R}^2$ with the map

$$p(z;t)=(t_1z^{(1)},t_1^bt_2z^{(2)},t_1^ez^{(3)}).$$

The rational monomials φ_j^{-1}, $j=1,2$ are $\varphi_1^{-1}(\tau)=\tau_1$, $\varphi_2^{-1}(\tau)=\dfrac{\tau_2}{\tau_1^b}$ and $\psi_3=\dfrac{\tau_3}{\tau_1^e}$. A cofinal family of $C(\xi)$ is $C(\xi,F^0)$ consisting of elements $S(W,\epsilon)$, $W=W_1\times W_2\times W_3$, defined as follows

$$S(W,\epsilon)=\left\{z\in X;\begin{array}{l}z^{(k)}\in W_k\quad(k=1,2,3),\\ |z^{(1)}|<\epsilon,\\ |z^{(2)}|<\epsilon|z^{(1)}|^b,\\ (|\xi^{(3)}|-\epsilon)|z^{(1)}|^e<|z^{(3)}|<(|\xi^{(3)}|+\epsilon)|z^{(1)}|^e\end{array}\right\},$$

where $\epsilon>0$ and each W_k is an $\mathbb{R}^+$-conic open subset in $\mathbb{C}^{m_k}$ containing the point $\xi^{(k)}$.

✓ **Case** $m=3, N=4$ **(b).** We have $Z_1=\{z^{(1)}=z^{(2)}=0\}$ and $Z_2=\{z^{(2)}=z^{(3)}=0\}$. Then we can define a normal deformation $\widetilde{X}=\mathbb{C}^n\times\mathbb{R}^2$ with the map $p(z;t)=(t_1z^{(1)},t_1^bt_2z^{(2)},t_2^ez^{(3)})$. The rational monomials φ_j^{-1}, $j=1,2$ are $\varphi_1^{-1}(\tau)=\tau_1$, $\varphi_2^{-1}(\tau)=\dfrac{\tau_2}{\tau_1^b}$ and $\psi_3=\dfrac{\tau_3\tau_1^{be}}{\tau_2^e}$. A cofinal family of $C(\xi)$ is $C(\xi,F^0)$ consisting of elements $S(W,\epsilon)$, $W=W_1\times W_2\times W_3$, defined as follows

$$S(W,\epsilon)=\left\{z\in X;\begin{array}{l}z^{(k)}\in W_k\quad(k=1,2,3),\\ |z^{(1)}|<\epsilon,\\ |z^{(2)}|<\epsilon|z^{(1)}|^b,\\ (|\xi^{(3)}|-\epsilon)|z^{(2)}|^e<|z^{(3)}||z^{(1)}|^{be}<(|\xi^{(3)}|+\epsilon)|z^{(2)}|^e\end{array}\right\},$$

where $\epsilon>0$ and each W_k is an $\mathbb{R}^+$-conic open subset in $\mathbb{C}^{m_k}$ containing the point $\xi^{(k)}$.

✓ **Case** $m=3, N=5$. In this case

$$A_\chi=\begin{pmatrix}1&b&e\\0&1&f\end{pmatrix}$$

with $b,e,f\in\mathbb{Q}_{>0}$, $bf\neq e$. We have $Z_1=\{0\}$ and $Z_2=\{z^{(2)}=z^{(3)}=0\}$. Then we can define a normal deformation $\widetilde{X}=\mathbb{C}^n\times\mathbb{R}^2$ with the map $p(z;t)=(t_1z^{(1)},t_1^bt_2z^{(2)},t_1^et_2^fz^{(3)})$. The rational monomials φ_j^{-1}, $j=1,2$ are $\varphi_1^{-1}(\tau)=\tau_1$,

$\varphi_2^{-1}(\tau) = \dfrac{\tau_2}{\tau_1^b}$ and $\psi_3 = \dfrac{\tau_3}{\tau_1^{e-bf}\tau_2^f}$. A cofinal family of $C(\xi)$ is $C(\xi, F^0)$ consisting of elements $S(W, \epsilon)$, $W = W_1 \times W_2 \times W_3$, defined as follows

$$S(W,\epsilon) = \left\{ z \in X; \begin{array}{l} z^{(k)} \in W_k \quad (k = 1,2,3), \\ |z^{(1)}| < \epsilon, \\ |z^{(2)}| < \epsilon |z^{(1)}|^b, \\ (|\xi^{(3)}| - \epsilon)|z^{(1)}|^{e-bf}|z^{(2)}|^f < |z^{(3)}| < (|\xi^{(3)}| + \epsilon)|z^{(1)}|^{e-bf}|z^{(2)}|^f \end{array} \right\},$$

where $\epsilon > 0$ and each W_k is an $\mathbb{R}^+$-conic open subset in $\mathbb{C}^{m_k}$ containing the point $\xi^{(k)}$.

✓ **Case** $m = 3, N = 6$**.** In this case

$$A_X = \begin{pmatrix} 1 & b & e \\ c & 1 & r \end{pmatrix}$$

with $b, c, e, r \in \mathbb{Q}_{>0}$, $bc \neq 1$, $br \neq e$, $ce \neq r$ and $1 - bc > 0$. We have $Z_1 = Z_2 = \{0\}$. Then we can define a normal deformation $\widetilde{X} = \mathbb{C}^n \times \mathbb{R}^2$ with the map $p(z;\, t) = (t_1 t_2^c z^{(1)}, t_1^b t_2 z^{(2)}, t_1^e t_2^r z^{(3)})$. Set $d = \dfrac{1}{1 - bc}$. The rational monomials φ_j^{-1}, $j = 1, 2$ are $\varphi_1^{-1}(\tau) = \left(\dfrac{\tau_1}{\tau_2^c}\right)^d$, $\varphi_2^{-1}(\tau) = \left(\dfrac{\tau_2}{\tau_1^b}\right)^d$ and $\psi_3 = \dfrac{\tau_3}{(\tau_1^{e-br}\tau_2^{r-ce})^d}$. A cofinal family of $C(\xi)$ is $C(\xi, F^0)$ consisting of elements $S(W, \epsilon)$, $W = W_1 \times W_2 \times W_3$, defined as follows

$$S(W,\epsilon) = \left\{ z \in X; \begin{array}{l} z^{(k)} \in W_k \quad (k = 1,2,3), \\ |z^{(1)}| < \epsilon^{\frac{1}{d}} |z^{(2)}|^c, \\ |z^{(2)}| < \epsilon^{\frac{1}{d}} |z^{(1)}|^b, \\ (|\xi^{(3)}| - \epsilon)(|z^{(1)}|^{e-br}|z^{(2)}|^{r-ce})^d < |z^{(3)}| \\ \qquad < (|\xi^{(3)}| + \epsilon)(|z^{(1)}|^{e-br}|z^{(2)}|^{r-ce})^d \end{array} \right\},$$

where $\epsilon > 0$ and each W_k is an $\mathbb{R}^+$-conic open subset in $\mathbb{C}^{m_k}$ containing the point $\xi^{(k)}$.

In the other cases one of the following occurs:

- the action is degenerate,
- we reduce to $m = 2$.

Appendix

In the Appendix we define conic objects and conic sheaves associated with multiple actions of $\mathbb{R}^+$. The proofs of the statements of Appendix A.1 and A.2 can be found in [9]. In Appendix A.3 we prove a decomposition theorem for open subanalytic subsets which is crucial for the computation of the sections of the multi-specialization.

A.1 Conic sheaves

Let k be a field. Let X be a real analytic manifold endowed with a subanalytic action μ of $\mathbb{R}^+$. In other words we have a subanalytic map

$$\mu : X \times \mathbb{R}^+ \to X,$$

which satisfies, for each $t_1, t_2 \in \mathbb{R}^+$:

$$\begin{cases} \mu(x, t_1 t_2) = \mu(\mu(x, t_1), t_2), \\ \mu(x, 1) = x. \end{cases}$$

Note that μ is open, in fact let $U \in \mathrm{Op}(X)$ and $(t_1, t_2) \in \mathrm{Op}(\mathbb{R}^+)$. Then

$$\mu(U, (t_1, t_2)) = \bigcup_{t \in (t_1, t_2)} \mu(U, t),$$

and $\mu(\cdot, t) : X \to X$ is a homeomorphism (with inverse $\mu(\cdot, t^{-1})$). We have a diagram

$$X \xrightarrow{j} X \times \mathbb{R}^+ \overset{\mu}{\underset{p}{\rightrightarrows}} X,$$

where $j(x) = (x, 1)$ and p denotes the projection. We have $\mu \circ j = p \circ j = \mathrm{id}$.

Definition A.1. *(i) Let S be a subset of X. We set $\mathbb{R}^+ S = \mu(S, \mathbb{R}^+)$. If $U \in \mathrm{Op}(X)$, then $\mathbb{R}^+ U \in \mathrm{Op}(X)$ since μ is open.*

(ii) Let S be a subset of X. We say that S is conic if $S = \mathbb{R}^+ S$. In other words, S is invariant by the action of μ.

(iii) An orbit of μ is the set $\mathbb{R}^+ x$ with $x \in X$.

We assume that the orbits of μ are contractible. For each $x \in X$ there are two possibilities: either $\mathbb{R}^+ x = x$ or $\mathbb{R}^+ x \simeq \mathbb{R}$.

Definition A.2. *We say that a subset S of X is $\mathbb{R}^+$-connected if $S \cap \mathbb{R}^+ x$ is connected for each $x \in S$.*

Lemma A.3. *(i) Let $S_1, S_2 \subset X$ and suppose that S_2 is conic. Then $\mathbb{R}^+(S_1 \cap S_2) = \mathbb{R}^+ S_1 \cap S_2$. (ii) If S_1 and S_2 are $\mathbb{R}^+$-connected, then $S_1 \cap S_2$ is $\mathbb{R}^+$-connected.*

Let X, Y topological spaces endowed with an action (μ_X and μ_Y respectively) of $\mathbb{R}^+$.

Definition A.4. *A continuous function $f : X \to Y$ is said to be conic if for each $x \in X$, $a \in \mathbb{R}^+$ we have $f(\mu_X(x,a)) = \mu_Y(f(x),a)$.*

Lemma A.5. *Let $f : X \to Y$ be a conic map. (i) Suppose that $S \subset Y$ is $\mathbb{R}^+$-connected (resp. conic). Then $f^{-1}(S)$ is $\mathbb{R}^+$-connected (resp. conic). (ii) Suppose that $Z \subset X$ is conic. Then $f(Z)$ is conic.*

Let X be a real analytic manifold endowed with a subanalytic action of $\mathbb{R}^+$. Denote by X_{sa} the associated subanalytic site.

Definition A.6. *A sheaf of k-modules F on X_{sa} is conic if the restriction morphism $\Gamma(\mathbb{R}^+U;F) \to \Gamma(U;F)$ is isomorphic for each $\mathbb{R}^+$-connected $U \in \mathrm{Op}^c(X_{sa})$ with $\mathbb{R}^+U \in \mathrm{Op}(X_{sa})$.*

(i) We denote by $\mathrm{Mod}_{\mathbb{R}^+}(k_{X_{sa}})$ the subcategory of $\mathrm{Mod}(k_{X_{sa}})$ consisting of conic sheaves.

(ii) We denote by $D^b_{\mathbb{R}^+}(k_{X_{sa}})$, the subcategory of $D^b(k_{X_{sa}})$ consisting of objects F such that $H^j(F)$ belongs to $\mathrm{Mod}_{\mathbb{R}^+}(k_{X_{sa}})$ for all $j \in \mathbb{Z}$.

Assume the hypothesis below:

$$\begin{cases} \text{(i) every } U \in \mathrm{Op}^c(X_{sa}) \text{ has a finite covering consisting} \\ \quad \text{of } \mathbb{R}^+\text{-connected subanalytic open subsets,} \\ \text{(ii) for any } U \in \mathrm{Op}^c(X_{sa}) \text{ we have } \mathbb{R}^+U \in \mathrm{Op}(X_{sa}), \\ \text{(iii) for any } x \in X \text{ the set } \mathbb{R}^+x \text{ is contractible,} \\ \text{(iv) there exists a covering } \{V_n\}_{n\in\mathbb{N}} \text{ of } X_{sa} \text{ such that} \\ \quad V_n \text{ is } \mathbb{R}^+\text{-connected and } V_n \subset\subset V_{n+1} \text{ for each } n. \end{cases} \tag{A.1}$$

The following result was proven in [21].

Proposition A.7. *Assume* (A.1). *Let $U \in \mathrm{Op}(X_{sa})$ be $\mathbb{R}^+$-connected and such that $\mathbb{R}^+U \in \mathrm{Op}(X_{sa})$. Let $F \in D^b_{\mathbb{R}^+}(k_{X_{sa}})$. Then*

$$R\Gamma(\mathbb{R}^+U;F) \xrightarrow{\sim} R\Gamma(U;F).$$

A.2 Multi-conic sheaves

Let X be a topological space with ℓ actions $\{\mu_i\}_{i=1}^{\ell}$ of $\mathbb{R}^+$ such that $\mu_i(\mu_j(x,t_j),t_i) = \mu_j(\mu_i(x,t_i),t_j)$. We have a map

$$\begin{aligned} \mu : X \times (\mathbb{R}^+)^\ell &\to X \\ (x,(t_1,\dots,t_\ell)) &\mapsto \mu_1(\cdots\mu_\ell(x,t_\ell),\dots,t_1). \end{aligned}$$

Definition A.8. *(i) Let S be a subset of X. We set $\mathbb{R}^+_i S = \mu_i(S,\mathbb{R}^+)$. If $U \in \mathrm{Op}(X)$, then $\mathbb{R}^+_i U \in \mathrm{Op}(X)$ since μ_i is open for each $i = 1,\dots,\ell$.*

(ii) Let S be a subset of X. Let $J = \{i_1, \dots, i_k\} \subseteq \{1, \dots, \ell\}$. We set

$$\mathbb{R}^+_J S = \mathbb{R}^+_{i_1} \cdots \mathbb{R}^+_{i_k} S = \mu_{i_1}(\cdots \mu_{i_k}(S, \mathbb{R}^+), \dots, \mathbb{R}^+), \quad i_1, \dots i_k \in J.$$

We set $(\mathbb{R}^+)^\ell S = \mathbb{R}^+_{\{1,\dots,\ell\}} S = \mu(S, (\mathbb{R}^+)^\ell)$. If $U \in \mathrm{Op}(X)$, then $\mathbb{R}^+_J U \in \mathrm{Op}(X)$ since μ_i is open for each $i \in \{1, \dots, \ell\}$.

(iii) Let S be a subset of X. We say that S is $(\mathbb{R}^+)^\ell$-conic if $S = (\mathbb{R}^+)^\ell S$. In other words, S is invariant by the action of μ_i, $i = 1, \dots, \ell$.

Definition A.9. *(i) We say that a subset S of X is $\mathbb{R}^+_i$-connected if $S \cap \mathbb{R}^+_i x$ is connected for each $x \in S$.*

(ii) We say that a subset S of X is $(\mathbb{R}^+)^\ell$-connected if there exists a permutation $\sigma : \{1, \dots, \ell\} \to \{1, \dots, \ell\}$ such that

$$\begin{cases} S \text{ is } \mathbb{R}^+_{\sigma(1)}\text{-connected}, \\ \mathbb{R}^+_{\sigma(1)} S \text{ is } \mathbb{R}^+_{\sigma(2)}\text{-connected}, \\ \quad \vdots \\ \mathbb{R}^+_{\sigma(1)} \cdots \mathbb{R}^+_{\sigma(\ell-1)} S \text{ is } \mathbb{R}^+_{\sigma(\ell)}\text{-connected}. \end{cases} \tag{A.2}$$

The following results follow from the case $\ell = 1$.

Lemma A.10. *(i) Let $S_1, S_2 \subset X$ and suppose that S_2 is $(\mathbb{R}^+)^\ell$-conic. Then $(\mathbb{R}^+)^\ell (S_1 \cap S_2) = (\mathbb{R}^+)^\ell S_1 \cap S_2$. (ii) If moreover S_1 is $(\mathbb{R}^+)^\ell$-connected then $S_1 \cap S_2$ is $(\mathbb{R}^+)^\ell$-connected.*

Remark A.11. *In (ii) of Lemma A.10 we have to assume that S_2 is $(\mathbb{R}^+)^\ell$-conic. Indeed it is not true that the intersection of two $(\mathbb{R}^+)^\ell$-connected is $(\mathbb{R}^+)^\ell$-connected in general.*

Let X, Y topological spaces endowed with ℓ actions $\{\mu_{Xi}\}_{i=1}^\ell$, $\{\mu_{Yi}\}_{i=1}^\ell$ of $\mathbb{R}^+$.

Definition A.12. *A continuous function $f : X \to Y$ is said to be $(\mathbb{R}^+)^\ell$-conic if for each $x \in X$, $a \in \mathbb{R}^+$ we have $f(\mu_{Xi}(x, a)) = \mu_{Yi}(f(x), a)$, $i = 1, \dots, \ell$.*

Lemma A.13. *Let $f : X \to Y$ be a $(\mathbb{R}^+)^\ell$-conic map. (i) Suppose that $S \subset Y$ is $(\mathbb{R}^+)^\ell$-connected (resp. $(\mathbb{R}^+)^\ell$-conic). Then $f^{-1}(S)$ is $(\mathbb{R}^+)^\ell$-connected (resp. $(\mathbb{R}^+)^\ell$-conic). (ii) Suppose that $Z \subset X$ is $(\mathbb{R}^+)^\ell$-conic. Then $f(Z)$ is $(\mathbb{R}^+)^\ell$-conic.*

Let X be a real analytic manifold and denote by X_{sa} the associated subanalytic site. Assume that X is endowed with ℓ subanalytic $\mathbb{R}^+$-actions $\mu_1, \dots, \mu_\ell$ commuting with each other.

Definition A.14. *A sheaf of k-modules F on X_{sa} is $(\mathbb{R}^+)^\ell$-conic if it is conic with respect to each μ_i.*

(i) We denote by $\mathrm{Mod}_{(\mathbb{R}^+)^\ell}(k_{X_{sa}})$ the subcategory of $\mathrm{Mod}(k_{X_{sa}})$ consisting of $(\mathbb{R}^+)^\ell$-conic sheaves.

(ii) We denote by $D^b_{(\mathbb{R}^+)^\ell}(k_{X_{sa}})$, the subcategory of $D^b(k_{X_{sa}})$ consisting of objects F such that $H^j(F)$ belongs to $\mathrm{Mod}_{(\mathbb{R}^+)^\ell}(k_{X_{sa}})$ for all $j \in \mathbb{Z}$.

Let us assume the following hypothesis

$$\begin{cases} \text{(i) the pair } (X,\mu_i) \text{ satisfies (A.1) for each } i=1,\dots,\ell, \\ \text{(ii) every } U \in \mathrm{Op}^c(X_{sa}) \text{ has a finite covering consisting} \\ \quad \text{of } (\mathbb{R}^+)^\ell\text{-connected subanalytic open subsets,} \\ \text{(iii) we have } \mathbb{R}^+_J U \in \mathrm{Op}(X_{sa}) \text{ for any } U \in \mathrm{Op}^c(X_{sa}) \\ \quad \text{and any } J \subset \{1,\dots,\ell\}. \end{cases} \tag{A.3}$$

In this situation the orbits of μ_i, $i = 1,\dots,\ell$ are either $\mathbb{R}^+x \simeq \mathbb{R}$ or $\mathbb{R}^+x = x$.

Proposition A.15. *Assume* (A.3). *Let $U \in \mathrm{Op}(X_{sa})$ be $(\mathbb{R}^+)^\ell$-connected. Let $F \in D^b_{(\mathbb{R}^+)^\ell}(k_{X_{sa}})$. Then*

$$R\Gamma((\mathbb{R}^+)^\ell U;F) \xrightarrow{\sim} R\Gamma(U;F).$$

If X satisfies (A.3) (i)-(iii), then it follows from Proposition A.15 that for $(\mathbb{R}^+)^\ell$-conic subanalytic sheaves it is enough to study the cohomology of the sections on $(\mathbb{R}^+)^\ell$-conic open subsets.

A.3 A decomposition theorem for subanalytic open sets

Let $X = \mathbb{R}^n$ with coordinates $x = (x_1,\dots,x_n)$ endowed with the actions $\mu_j : X \times \mathbb{R}^+ \to X$, $j = 1,\dots,\ell$ defined by

$$\mu_j(x,\lambda) = \mu_j(x_1,\dots,x_n,\lambda) = (\lambda^{\alpha_{j1}}x_1,\dots,\lambda^{\alpha_{jn}},x_n),$$

with $\lambda \geq 0$ and $\alpha_{ji} \in \mathbb{R}$, $\alpha_{ji} \neq 0$ if $i \in I_j \subseteq \{1,\dots,n\}$. We can represent the actions μ_j, $j = 1,\dots,\ell$ in a matrix form as follows: let $x = (x_1,\dots,x_n)$, $\lambda = (\lambda_1,\dots,\lambda_\ell)$, $A = (\alpha_{ji}) \in M_{\ell,n}(\mathbb{R})$. Then

$$\mu(x,\lambda) = \mu_1(\cdots(\mu_\ell(x,\lambda_\ell)\cdots),\lambda_1) = xe^{\log\lambda A} \tag{A.4}$$

where $\log\lambda = (\log\lambda_1,\dots,\log\lambda_\ell)$ and, given $y = (y_1,\dots,y_n)$, $e^y = \delta_{ij}e^{y_i} \in M_{n,n}(\mathbb{R})$ (δ_{ij} is the Kronecker's delta) is a diagonal matrix. So the i-th diagonal entry of $e^{\log\lambda A}$ is $\prod_{j=1}^{\ell}\lambda^{\alpha_{ji}}$. We consider the equivalence classes $\hat{I}_k$ defined as follows:

$$k_1, k_2 \in \hat{I}_k \text{ iff } \pi_{k_1}(\mu_j(x,\lambda)) = \pi_{k_2}(\mu_j(x,\lambda)) \text{ for any } j \in 1,\dots,\ell \text{ and any } \lambda \geq 0.$$

Here π_{k_i} denotes the projection on the k_i-th coordinate, $i = 1,2$. That means $\alpha_{jk_1} = \alpha_{jk_2}$ for each $j = 1,\dots,\ell$. So we may assume that $\{1,\dots,n\} = \bigsqcup_{k=1}^{N}\hat{I}_k$.

Let $m_k = \sharp \hat{I}_k$. Then $\sum_{k=1}^N m_k = n$. Consider the morphism of manifolds

$$\begin{aligned} \varphi : \widetilde{X}' = \times_{k=1}^N \mathbb{S}^{m_k - 1} \times \mathbb{R}^N &\to \mathbb{R}^n \\ (\vartheta_k, r_k)_{k=1}^N &\mapsto (r_k \iota_k(\vartheta_k))_{k=1}^N, \end{aligned} \tag{A.5}$$

where $\iota_k : \mathbb{S}^{m_k-1} \hookrightarrow \mathbb{R}^{m_k}$ denotes the embedding. Endow $\times_{k=1}^N \mathbb{S}^{m_k-1} \times \mathbb{R}^N$ with the actions $\widetilde{\mu}'_j = \mu_j$, $j = 1, \ldots, \ell - 1$, $\widetilde{\mu}'_j((\vartheta_k, r_k)_{k=1}^N, \lambda) = (\vartheta_k, \lambda^{\alpha_{jk}} r_k)_{k=1}^N$. Then φ is a $(\mathbb{R}^+)^\ell$-conic map.

Let V be an $(\mathbb{R}^+)^\ell$-conic subanalytic subset in X. Let $S_k = \{x_i = 0;\, i \in \hat{I}_k\}$, and let $\pi_j : X \to S_k$ be the projection. We introduce the following conditions Va. and Vb. of V for each $k = 1, \ldots, N$.

Va. V does not intersect S_k.

Vb. $\pi_j(V) \subset \pi_j(V \cap S_k)$.

Proposition A.16. *Let V be a $(\mathbb{R}^+)^\ell$-conic subanalytic subset in X and let W be an open subanalytic neighborhood of V. If V satisfies the condition either Va. or Vb. for each j, then there exists a subanalytic subset W' in X satisfying the following conditions.*

1. *W' is an open neighborhood of V and contained in W.*
2. *W' is $(\mathbb{R}^+)^\ell$-connected.*
3. *$\mathbb{R}_1^+ \ldots \mathbb{R}_k^+ W'$ is also subanalytic in X for any $1 \le k \le \ell$.*

Proof. Thanks to the morphism of manifolds (A.5) we may reduce to $X = [0, +\infty]^N$ setting $\widetilde{\mu}'_j = \mu_j$ for short and omitting the variables ϑ_j (they are irrelevant being fixed by each μ_j, $j = 1, \ldots, \ell$). We argue by induction on the number of actions. We prove the assertion in several steps.

(a) Suppose that there exists $i \in I_\ell$ such that V satisfies Va. on i. Up to a permutation of coordinates, we may assume $i = N$. First, on $\{x_N \neq 0\}$ consider the homeomorphism ψ making the orbits of μ_ℓ orthogonal to $\{x_N = 1\}$. Namely, if $\mu_\ell(x_1, \ldots, x_{N-1}, x_N, \lambda) = (\lambda^{\alpha_{\ell 1}} x_1, \ldots, \lambda^{\alpha_{\ell N-1}} x_{N-1}, \lambda^{\alpha_{\ell N}} x_N)$ we set $\psi(x_1, \ldots, x_{N-1}, x_N) = (x_1 x_N^{-\alpha_{\ell 1}/\alpha_{\ell N}}, \ldots, x_{N-1} x_N^{-\alpha_{\ell N-1}/\alpha_{\ell N}}, x_N^{1/\alpha_{\ell N}})$. Then $\psi \circ \mu_\ell \circ (\psi^{-1} \times \mathrm{id}_{\mathbb{R}^+})$ acts only on the variable x_N. This corresponds to the change of coordinates sending the matrix (representing the action as in (A.4))

$$\begin{pmatrix} A & B \\ C & d \end{pmatrix} \quad \text{to} \quad \begin{pmatrix} A' & B' \\ 0 & 1 \end{pmatrix}$$

where A (resp. A') is a $(N-1) \times (N-1)$ matrix, B (resp. B') is a $(N-1) \times 1$ matrix, C (resp. 0) is a $1 \times (N-1)$ matrix and $d \neq 0$.

Replace V with $\psi(V)$ and μ_j with $\psi \circ \mu_j \circ (\psi^{-1} \times \mathrm{id}_{\mathbb{R}^+})$, $j = 1, \dots, \ell$. One can check easily that $V \cap \{x_N = 1\}$ (remark that $\psi(\{x_N = 1\}) = \{x_N = 1\}$) is conic with respect to the actions $\widetilde{\mu}_j$ defined as

$$\widetilde{\mu}_j(x_1, \dots, x_{N-1}, 1, \lambda) = (\lambda^{\alpha_{j1}} x_1, \dots, \lambda^{\alpha_{jN-1}} x_{N-1}, 1).$$

By the induction hypothesis $W \cap \{x_N = 1\}$ contains a subanalytic open neighborhood W_1 of $V \cap \{x_N = 1\}$ in $\{x_N = 1\}$ which is $(\widetilde{\mathbb{R}^+})^{\ell-1}$-connected (here $\widetilde{\mathbb{R}^+}$ means with respect to $\widetilde{\mu}_j$, $j = 1, \dots, \ell - 1$).

Let $\pi : (x_1, \dots, x_{N-1}, x_N) \mapsto (x_1, \dots, x_{N-1})$ be the projection. Then $W \cap \pi^{-1}(\pi(p))$ is a disjoint union of intervals. Let us consider the interval $(m(p), M(p))$ containing $x_N = 1$. Set

$$\widetilde{W} = \{(x_1, \dots, x_{N-1}, x_N);\ (x_1, \dots, x_{N-1}) \in \pi(W_1),\ m(p) < x_N < M(p)\}.$$

By construction $\widetilde{W}$ is subanalytic and $\mathbb{R}^+_\ell$-connected. Let us prove that it is $(\mathbb{R}^+)^\ell$-connected. Remark that $\mathbb{R}^+_j \mathbb{R}^+_\ell S = \mathbb{R}^+_\ell \widetilde{\mathbb{R}^+_j} S$ for each $j = 1, \dots, \ell - 1$ and $S \subseteq \{x_N = 1\}$ (here $\widetilde{\mathbb{R}^+_j}$ means conic with respect to $\widetilde{\mu}_j$). So by the induction hypothesis we are reduced to prove that $\mathbb{R}^+_\ell S$ is $\mathbb{R}^+_j$-connected if S is $\widetilde{\mathbb{R}^+_j}$-connected, $j = 1, \dots, \ell - 1$.

Let $x = (x_1, \dots, x_{N-1}, x_N), \mu_j(x, a) \in \mathbb{R}^+_\ell S$. We shall prove that $\mu_j(x, b) \in \mathbb{R}^+_\ell S$ if $b \in [1, a]$ (we assume without loss of generality that $a > 1$). We have $(x_1, \dots, x_{N-1}, 1) = \mu_\ell(x, x_N^{-1}), \widetilde{\mu}_j(x_1, \dots, x_{N-1}, 1, a) = \mu_\ell(\mu_j(x, a), (a^{\alpha_{jN}} x_N)^{-1}) \in S$. Since S is $\widetilde{\mathbb{R}^+_j}$-connected we have $\widetilde{\mu}_j(x_1, \dots, x_{N-1}, 1, b) \in S$ and hence $\mu_j(x, b) = \mu_\ell(\widetilde{\mu}_j(x_1, \dots, x_{N-1}, 1, b), b^{\alpha_{jN}} x_N) \in \mathbb{R}^+_\ell S$.

(b) Suppose that V satisfies Vb. for each $i \in I_\ell$. Up to shrink W, we may assume that W satisfies Vb. for each $i \in I_\ell$ as well. Let $J^+ = \{i \in I_\ell,\ \alpha_{\ell i} > 0\}$, $J^- = \{i \in I_\ell,\ \alpha_{\ell i} < 0\}$.

We first shrink W on $X \setminus (\{x_i = 0,\ i \in J^+\} \sqcup \{x_i = 0,\ i \in J^-\})$. First of all we consider the homeomorphism such that $x_i \mapsto x_i^{1/|\alpha_{\ell i}|}$ if $i \in J^- \sqcup J^+$ and $x_i \mapsto x_i$ otherwise. Then we may assume that $\alpha_{\ell i} = \pm 1$ when $\alpha_{\ell i} \neq 0$. Let $x \in X$ and set $|x|_+ = \left(\sum_{i \in J^+} x_i^2\right)^{1/2}, |x|_- = \left(\sum_{i \in J^-} x_i^2\right)^{1/2}$. Set $S = \{x \in X;\ |x|_+ = |x|_-\}$. One can check easily that with this definition the intersection of S with an orbit of μ_ℓ is a point (namely, given $x_0 \in X$ we have $\mathbb{R}^+_\ell x \cap S = \{\mu_\ell(x_0, \lambda)\}$ with $\lambda = \sqrt{|x_0|_- / |x_0|_+}$). Up to take a permutation of coordinates, we may assume that $N \in J^+$. Let us first assume that $x_N \neq 0$. As in (a), we may choose an homeomorphism ψ such that $\psi \circ \mu_\ell \circ (\psi^{-1} \times \mathrm{id}_{\mathbb{R}^+})$ acts only on the variable x_N.

Replace V with $\psi(V)$ and μ_j with $\psi \circ \mu_j \circ (\psi^{-1} \times \mathrm{id}_{\mathbb{R}^+})$, $j = 1, \dots, \ell$. One can check easily that $V \cap S$ is conic with respect to the actions $\widetilde{\mu}_j$ defined as $\widetilde{\mu}_j(x_1, \dots, x_{N-1}, x_N, \lambda) = \eta(\lambda^{\alpha_{j1}} x_1, \dots, \lambda^{\alpha_{jN-1}} x_{N-1})$, where $\eta : [0, +\infty]^{N-1} \xrightarrow{\sim} S$.

By the induction hypothesis $W \cap S$ contains a subanalytic open neighborhood W_1 of $V \cap S$ in S which is $(\widetilde{\mathbb{R}^+})^{\ell-1}$-connected (here $\widetilde{\mathbb{R}^+}$ means conic with respect to $\widetilde{\mu}_j$, $j = 1, \ldots, \ell - 1$).

Let $\pi : (x_1, \ldots, x_{N-1}, x_N) \mapsto (x_1, \ldots, x_{N-1})$ be the projection. Then $W \cap \pi^{-1}(\pi(p))$ is a disjoint union of intervals. Let us consider the interval $(m(p), M(p))$ intersecting S. Set

$$\widetilde{W}_1' = \{(x_1, \ldots, x_{N-1}, x_N);\ (x_1, \ldots, x_{N-1}) \in \pi(W_1),\ m(p) < x_N < M(p)\}.$$

By construction $\widetilde{W}_1'$ is subanalytic and $\mathbb{R}^+_\ell$-connected. To show that it is $(\mathbb{R}^+)^\ell$-connected one can easily adapt the proof of (a).

Now let us consider $W \cap \{x_N = 0\}$. Up to shrink W, we may assume that $\widetilde{W}_1' \cup (W \cap \{x_N = 0\})$ is open. By induction on the dimension of J^+ we may construct $\widetilde{W}_1^0 \subset W \cap \{x_N = 0\}$ with the required properties. By construction $\widetilde{W}_1 = \widetilde{W}_1' \cup \widetilde{W}_1^0$ is subanalytic and $\mathbb{R}^+_\ell$-connected.

Let us now shrink $W \cap \{x_i = 0,\ i \in J^+\}$. The same method will apply to $W \cap \{x_i = 0,\ i \in J^-\}$. The set $V \cap \{x_i = 0,\ i \in J^+\} \cap \{x_i = 0,\ i \in J^-\}$ is $\mathbb{R}^+_j$-conic, $j = 1, \ldots, \ell - 1$ (the action μ_ℓ is trivial there). By the induction hypothesis we may assume that $W \cap \{x_i = 0,\ i \in J^+\} \cap \{x_i = 0,\ i \in J^-\}$ is $(\mathbb{R}^+)^{\ell-1}$-connected. The intersection of the orbits of μ_ℓ and W are homeomorphic to a disjoint union of intervals. Let us choose the ones whose boundary intersects $\{x_i = 0,\ i \in J^-\}$. Let us prove that this provides a set satisfying the hypothesis.

Assume that $N \in J^-$ and $x_N \neq 0$. As in (a), we may choose an homeomorphism ψ such that $\psi \circ \mu_\ell \circ (\psi^{-1} \times \mathrm{id}_{\mathbb{R}^+})$ acts only on the variable x_N. Replace V with $\psi(V)$ and μ_j with $\psi \circ \mu_j \circ (\psi^{-1} \times \mathrm{id}_{\mathbb{R}^+})$, $j = 1, \ldots, \ell$. Let $\pi : (x_1, \ldots, x_{N-1}, x_N) \mapsto (x_1, \ldots, x_{N-1})$ be the projection. One can check easily that $\pi(V)$ is conic with respect to the actions $\widetilde{\mu}_j$ defined as $\widetilde{\mu}_j(x_1, \ldots, x_{N-1}, \lambda) = (\lambda^{\alpha_{j1}} x_1, \ldots, \lambda^{\alpha_{jN-1}} x_{N-1})$ and that $\pi(W)$ is $(\widetilde{\mathbb{R}^+})^{\ell-1}$-connected (here $\widetilde{\mathbb{R}^+}$ means with respect to $\widetilde{\mu}_j$, $j = 1, \ldots, \ell - 1$). This is because $\psi^{-1}(W) \cap \{x_i = 0,\ i \in J^+\} \cap \{x_i = 0,\ i \in J^-\}$ is $(\mathbb{R}^+)^{\ell-1}$-connected and $\psi^{-1}(W)$ satisfies Vb.

Let $p \in W$. Then $W \cap \pi^{-1}(\pi(p))$ is a disjoint union of intervals. Let us consider the interval $(m(p), M(p))$ with $m(p) = 0$. Set

$$\widetilde{W}_0^+ = \{(x_1, \ldots, x_{N-1}, x_N);\ (x_1, \ldots, x_{N-1}) \in \pi(W),\ 0 < x_N < M(p)\}.$$

By construction $\widetilde{W}_0^+$ is subanalytic and $\mathbb{R}^+_\ell$-connected. To show that it is $(\mathbb{R}^+)^\ell$-connected one can easily adapt the proof of (a). On $x_N = 0$ the proof is similar.

To end the proof, let us show that $\widetilde{W} = \widetilde{W}_1 \cup \widetilde{W}_0^+ \cup \widetilde{W}_0^-$ is a neighborhood of $\widetilde{W}_0^+ \cup \widetilde{W}_0^+$. We argue by contradiction. We suppose that there exists $y \in \widetilde{W}_0^+ \cup \widetilde{W}_0^+$

and that for each $\varepsilon > 0$ there is $y_\varepsilon \in \{x \in X;\ |x-y| < \varepsilon,\ |x|_+, |x|_- > 0\}$ with $y_\varepsilon \notin \widetilde{W}$. Suppose that $y \in \widetilde{W}_0^-$ (if $y \in \widetilde{W}_0^+$ the proof is similar). Taking ε small enough, we may assume that $\mathbb{R}_\ell^+ y_\varepsilon \cap S = \mu_\ell(y_\varepsilon, \lambda_\varepsilon)$ with $\lambda_\varepsilon \in \mathbb{R}^+$. By construction of $\widetilde{W}$ there must be $z_\varepsilon \in \mu_\ell(y_\varepsilon, [\lambda_\varepsilon, 1])$ (or $z_\varepsilon \in \mu(y_\varepsilon, [1, \lambda_\varepsilon])$) with $z_\varepsilon \notin W$.

We may also assume that for ε small enough, z_ε belongs to $K \subset W$ with K compact. The set K is constructed as follows:
- if $y \notin \widetilde{W}_0^+$, take a compact neighborhood $B = \{x \in X;\ |x-y| \le \delta\}$ of y contained in W and set $K = \mathbb{R}_\ell^+ B \cap \{|x|_- \le |y|_- + \delta\} \cap \{|x|_+ \le |x|_-\}$,
- if $y \in \widetilde{W}_0^- \cap \widetilde{W}_0^+$, take a compact neighborhood $B = \{x \in X;\ |x-y| \le \delta\}$ of y contained in W and set $K = \mathbb{R}_\ell^+ B \cap \{|x|_- \le \delta\} \cap \{|x|_+ \le \delta\}$.
Being a neighborhood of $\widetilde{W}_0^+ \cup \widetilde{W}_0^+$, W must contain such a K for δ small enough.

Then we can extract a sequence z_n with

$$z_n \in \mu_\ell(y_{\varepsilon_n}, [\lambda_{\varepsilon_n}, 1]) \text{ or } z_n \in \mu_\ell(y_{\varepsilon_n}, [1, \lambda_{\varepsilon_n}])$$

which has a limit z in K. This leads to a contradiction since $z \in W$ and W is open.

□

A.4 A proof for Proposition 1.2

In this appendix, we give the proof for Proposition 1.2. Clearly ii. of the proposition implies i. of the proposition. We will show the converse implication. We may assume $q = 0$ and $(T^*X)_q = \mathbb{R}^n$. Furthermore, by a linear coordinates transformation, we also assume that each V_k in the definition is the vector subspace spanned by the vectors dx_i $(i \in E_k)$ for some $E_k \subset \{1, \ldots, n\}$. We will prove the claim by the induction with respect to n, i.e., the dimension of X. If $n = 0$, then the claim clearly holds. Suppose that the claim were true for $0, 1, \ldots, n-1$ and we will show the claim for $n > 0$. We need the following lemma.

Lemma A.17. *Let $\chi = \{M_1, \ldots, M_\ell\}$ be a family of closed submanifolds in X, and let g_j $(j = 1, \ldots, \ell)$ be a real analytic function on M_j satisfying $g_{j_1} = g_{j_2}$ on $M_{j_1} \cap M_{j_2}$ for any $1 \le j_1, j_2 \le \ell$. Suppose Proposition 1.2 were true and χ is simultaneously linearizable at $q \in M$. Then there exists a real analytic function $g(x)$ in an open neighborhood of q such that $g = g_j$ on M_j. Furthermore, if $d_{M_j} g_j(q) = 0$ holds for $j = 1, \ldots, \ell$, then we can choose such a $g(x)$ with $dg(q) = 0$.*

Proof. We may assume $q = 0$ and $X = \mathbb{R}^n$. We choose a system of coordinates $(x_1, \ldots, x_n)$ satisfying that each M_j is given by $\{x \in \mathbb{R}^n;\ x_i = 0\ (i \in I_j)\}$ for some $I_j \subset \{1, \ldots, n\}$. Let π_j be the orthogonal projection from $\mathbb{R}^n$ to M_j. For a subset $\alpha = \{j_1, \ldots, j_p\} \subset \{1, \ldots, \ell\}$, we set $M_\alpha = M_{j_1} \cap \cdots \cap M_{j_p}$ and define the function g_α in M_α by $g_j|_{M_\alpha}$ for some $j \in \alpha$. Note that, by the gluing conditions, g_α is independent of the choice of $j \in \alpha$. Let π_α denote the orthogonal projection from $\mathbb{R}^n$ to M_α. Then

$$g(x) = \sum_{\alpha \neq \emptyset} (-1)^{|\alpha|+1} g_\alpha(\pi_\alpha(x))$$

gives a desired real analytic function where α ranges through non-empty subsets of $\{1, \dots, \ell\}$ and $|\alpha|$ denotes the number of elements of α. □

Set $X_k := \{x \in \mathbb{R}^n;\, x_i = 0\ (i \in E_k)\}$. We denote by $\pi_k : \mathbb{R}^n \to X_k$ be the orthogonal projection. Define $\mathcal{J}_k := \{j \in \{1, \dots, \ell\};\, V_k \subset (T^*_{M_j} X)_q\}$. Note that, in a neighborhood of 0, π_k gives an isomorphism between M_j and $\pi_k(M_j) \subset X_k$ for any $j \in \mathcal{J}_k$.

Now let us consider a $1 \le k \le m$ for which $\mathcal{J}_k$ is non-empty. Suppose that $k = 1$ is such a k. For each $j \in \mathcal{J}_1$ and $i \in E_1$, because of $dx_i \in (T^*_{M_j} X)_0$, we can find a real analytic function $g_{j,i}$ in an open neighborhood of 0 with $g_{j,i} = 0$ on M_j and $dg_{j,i}(0) = dx_i$. Then, by the implicit function theorem, we may assume that each $g_{j,i}$ has a form $x_i - g'_{j,i}(x')$ where x' denotes the coordinates x_i with $i \in \{1, \dots, n\} \setminus E_1$ (i.e., the coordinates of X_1) and $g'_{j,i}(x')$ is a real analytic function in X_1 with $g'_{j,i}(0) = 0$ and $d_{X_1} g'_{j,i}(0) = 0$.

Let us consider the family $\chi' = \{\pi_1(M_j)\}_{j \in \mathcal{J}_1}$ of closed submanifolds in X_1. Clearly $(T^*X_1)_0$ contains the vector subspace $V_2 \oplus \cdots \oplus V_m$ such that each $(T^*_{\pi_1(M_j)} X)$ is a part of this direct sum for $j \in \mathcal{J}_1$.

Lemma A.18. *For any j_1 and j_2 in $\mathcal{J}_1$, the subset $\pi_1(M_{j_1}) \cap \pi_1(M_{j_2})$ is a closed submanifold in X_1 and we have*

$$(T^*_{\pi_1(M_{j_1})} X_1)_0 + (T^*_{\pi_1(M_{j_2})} X_1)_0 = (T^*_{\pi_1(M_{j_1}) \cap \pi_1(M_{j_2})} X_1)_0.$$

Proof. We get

$$\kappa := \dim (M_{j_1} \cap M_{j_2}) = \dim X - \dim ((T^*_{M_{j_1}} X)_0 + (T^*_{M_{j_2}} X)_0) \tag{A.6}$$

and

$$\begin{aligned} &\dim X - \dim ((T^*_{M_{j_1}} X)_0 + (T^*_{M_{j_2}} X)_0) \\ &= \dim X_1 - \dim ((T^*_{\pi_1(M_{j_1})} X_1)_0 + (T^*_{\pi_1(M_{j_2})} X_1)_0) \\ &\ge \dim (\pi_1(M_{j_1}) \cap \pi_1(M_{j_2})). \end{aligned} \tag{A.7}$$

Here the last dimension denotes that of an analytic germ at 0. As a result, we obtain $\kappa \ge \dim (\pi_1(M_{j_1}) \cap \pi_1(M_{j_2}))$. Let L be an irreducible component of the germ of $\pi_1(M_{j_1}) \cap \pi_1(M_{j_2})$. Note that $\dim L \le \kappa$ also holds. Then, since $M_{j_1} \cap M_{j_2}$ is defined by

$$\left\{ (\{g'_{j_1,i}(x')\}_{i \in E_1}, x') \in X;\ \begin{matrix} g'_{j_1,i}(x') = g'_{j_2,i}(x')\ (i \in E_1), \\ x' \in \pi_1(M_{j_1}) \cap \pi_1(M_{j_2}) \end{matrix} \right\},$$

if either $\dim L < \kappa$ or $g'_{j_1,i} \neq g'_{j_2,i}$ on L for some $i \in E_1$ holds, then we can find a point q' such that the dimension of the germ of $M_{j_1} \cap M_{j_2}$ at q' is strictly less than κ, which contradicts the fact that $M_{j_1} \cap M_{j_2}$ is smooth and its dimension is κ. Hence we can conclude that every irreducible components of $\pi_1(M_{j_1}) \cap \pi_1(M_{j_2})$ has dimension κ and we also have $g'_{j_1,i} = g'_{j_2,i}$ on $\pi_1(M_{j_1}) \cap \pi_1(M_{j_2})$ for any $i \in E_1$. Therefore we have

$$M_{j_1} \cap M_{j_2} = \left\{(\{g'_{j_1,i}(x')\}_{i \in E_1}, x') \in X;\ x' \in \pi_1(M_{j_1}) \cap \pi_1(M_{j_2})\right\},$$

from which $\pi_1(M_{j_1} \cap M_{j_2}) = \pi_1(M_{j_1}) \cap \pi_1(M_{j_2})$ follows. Clearly $\pi_1(M_{j_1} \cap M_{j_2})$ is a closed submanifold and its dimension is still κ. Therefore $\pi_1(M_{j_1}) \cap \pi_1(M_{j_2})$ is a closed submanifold in X_1. Furthermore, by(A.6) and (A.7), we get

$$\begin{aligned} &\dim\left((T^*_{\pi_1(M_{j_1})}X_1)_0 + (T^*_{\pi_1(M_{j_2})}X_1)_0\right) \\ &= \dim X_1 - \kappa = \dim (T^*_{\pi_1(M_{j_1}) \cap \pi_1(M_{j_2})}X_1)_0. \end{aligned}$$

Hence we have obtained the last equation in the lemma. □

By these observations, the family $\chi' = \{\pi_1(M_j)\}_{j \in \mathcal{J}_1}$ of closed submanifolds in X_1 is simultaneously linearizable. Hence, by the induction hypothesis, the claim ii. of the proposition for χ' holds. It follows from the proof of the above lemma that, for any j_1, j_2 in $\mathcal{J}_1$ and $i \in E_1$, we have $g'_{j_1,i} = g'_{j_2,i}$ on $\pi_1(M_{j_1}) \cap \pi_1(M_{j_2})$. Hence, for each $i \in E_1$, by applying Lemma A.17 to functions $g'_{j,i}$ on $\pi_1(M_j)$ ($j \in \mathcal{J}_1$), there exists a real analytic function $g'_i(x')$ in X_1 such that $g'_i = g'_{j,i}$ on $\pi_1(M_j)$ for $j \in \mathcal{J}_1$ and $d_{X_1} g'_i(0) = 0$. Thus we have obtained the real analytic function $f_i(x) := x_i - g'_i(x')$ ($i \in E_1$) such that $df_i(0) = dx_i$ and $f_i(x) = 0$ on M_j ($j \in \mathcal{J}_1$). By repeating this procedure for $E_2, \ldots, E_m$ respectively, we can obtain n-real analytic functions f_i ($i = 1, \ldots, n$). Here, if $\mathcal{J}_k$ is empty, then we simply set $f_i(x) := x_i$ ($i \in E_k$). It follows from the construction that these f_i's form a system of local coordinates of X near 0 and each M_j is defined by equations $f_i = 0$ ($i \in \bigcup_{V_k \subset (T^*_{M_j}X)_q} E_k$). Hence we have shown the claim for n and this completes the proof.

Bibliography

[1] E. Bierstone, D. Milman, Semianalytic and subanalytic sets, Publ. I.H.E.S. **67**, pp. 5-42 (1988).

[2] V. Colin, Specialization of formal cohomology and asymptotic expansions, Publ. RIMS, Kyoto Univ. **37** pp. 37-69 (2001).

[3] M. Coste, An introduction to o-minimal geometry, Dip. Mat. Univ. Pisa, Dottorato di Ricerca in Matematica, Istituti Editoriali e Poligrafici Internazionali, Pisa (2000).

[4] A. D' Agnolo, M. Kashiwara, Riemann-Hilbert correspondence for holonomic $\mathcal{D}$-modules, Publ. Math. Inst. Hautes Études Sci., **123**, pp. 69-197 (2016).

[5] J. M. Delort, Microlocalisation simultanée et problème de Cauchy ramifié, Compositio Math, **100**, pp. 171-204 (1996).

[6] A. Dufresnoy, Sur l'oprateur d" et les fonctions diffrentiables au sens de Whitney, Ann. Inst. Fourier, Grenoble, **29** N. **1** pp. 229-238 (1979).

[7] F. Galindo, J. Sanz, On strongly asymptotically developable functions and the Borel-Ritt theorem, Studia Math. **133** N. **3** pp. 231-248 (1999).

[8] N. Honda, L. Prelli, S. Yamazaki, Multi-microlocalization and microsupport, Bulletin Soc. Math. France, **144** N. 3 pp. 569-611 (2016).

[9] N. Honda, L. Prelli, Multi-specialization and multi-asymptotic expansions, Advances in Math. **232**, pp. 432-498 (2013).

[10] N. Honda, L. Prelli, Generalization of multi-specializations and multi-asymptotics, arXiv:1507.04572.

[11] N. Honda, L. Prelli, Multi-asymtotic expansions, to appear.

[12] N. Honda, L. Prelli, Direct and inverse image formulas for multi-specialization, to appear.

[13] L. Hörmander, An introduction to complex analysis in several variables, North-Holland (1973).

[14] M. Kashiwara, P. Schapira, Sheaves on manifolds, Grundlehren der Math. **292**, Springer-Verlag, Berlin (1990).

[15] M. Kashiwara, P. Schapira, Moderate and formal cohomology associated with constructible sheaves, Mémoires Soc. Math. France **64** (1996).

[16] M. Kashiwara, P. Schapira, Ind-sheaves, Astérisque **271** (2001).

[17] S. Łojaciewicz, Sur le problème de la division, Studia Mathematica **8**, (1959) pp. 87-136.

[18] H. Majima, Asymptotic Analysis for Integrable Connections with Irregular Singular Points, Lecture Notes in Math. **1075**, Springer-Verlag, Berlin (1984).

[19] B. Malgrange, Ideals of differentiable functions, Tata Institute, Oxford University Press (1966).

[20] L. Prelli, Sheaves on subanalytic sites, Rend. Sem. Mat. Univ. Padova Vol. **120**, pp. 167-216 (2008).

[21] L. Prelli, Conic sheaves on subanalytic sites and Laplace transform, Rend. Sem. Mat. Univ. Padova Vol. **125**, pp. 173-206 (2011).

[22] L. Prelli, Microlocalization of subanalytic sheaves, Mémoires Soc. Math. France **135** (2013).

[23] P. Schapira, Vanishing of temperate cohomology on complex manifolds, Annales Henri Lebesugue. **4**, pp. 863-877 (2021).

[24] P. Schapira, K. Takeuchi, Déformation normale et bispécialisation, C. R. Acad. Sci. Paris Math. **319**, pp. 707-712 (1994).

[25] K. Takeuchi, Binormal deformation and bimicrolocalization, Publ. RIMS, Kyoto Univ. **32** pp. 277-322 (1996).

[26] L. Van Den Dries, Tame topology and o-minimal structures, London Math. Society Lecture Notes Series **248**, Cambridge University Press, Cambridge (1998).

[27] H. Whitney, Functions differentiable on the boundaries of regions, Ann. Math. **35** pp. 482-485 (1934).